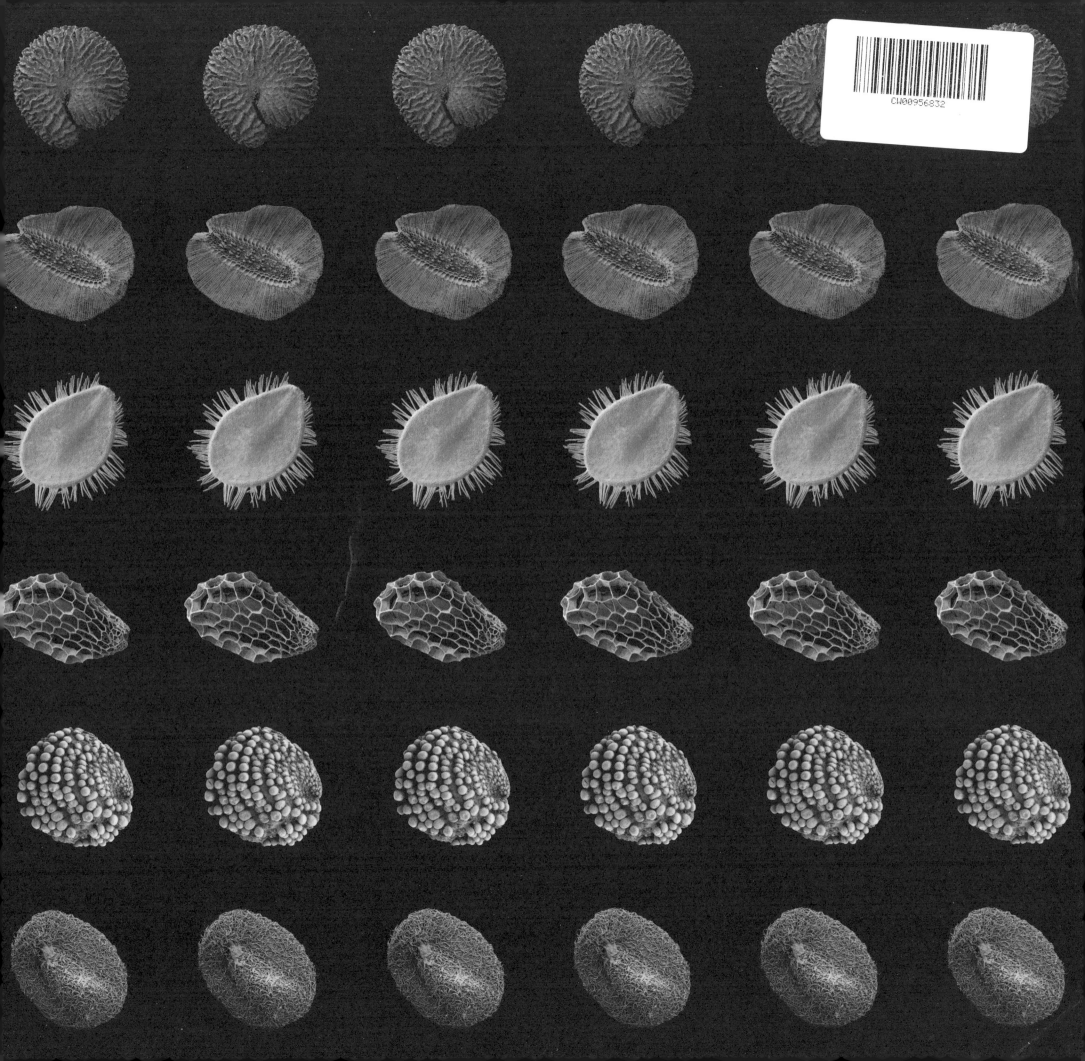

With love from
Annie and Chris.
Christmas 2007

SEEDS

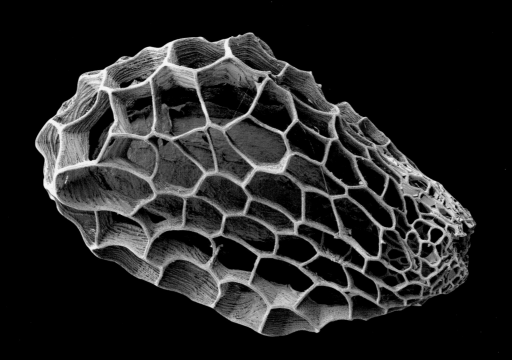

SEEDS

TIME CAPSULES OF LIFE

Rob Kesseler & Wolfgang Stuppy

EDITED BY ALEXANDRA PAPADAKIS

PREFACE BY HRH THE PRINCE OF WALES

FIREFLY BOOKS

A FIREFLY BOOK

Published by Firefly Books Ltd. 2006

Copyright © 2006 Rob Kesseler, Wolfgang Stuppy and Papadakis Publisher

Rob Kesseler and Wolfgang Stuppy hereby assert their moral right to be identified as authors of this work.

All rights reserved. No part of this publication may be reproduced, stored in a retrieval system, or transmitted in any form or by any means, electronic, mechanical, photocopying, recording or otherwise, without the prior written permission of the Publisher.

First Printing

Publisher Cataloging-in-Publication Data (U.S.)

Kesseler, Rob.
Seeds : time capsules of life / Rob Kesseler, & Wolfgang Stuppy ; edited by Alexandra Papadakis.
[264] p. : 15 col. ill. ; 252 col. photos. ; 30 x 24 cm.
Includes bibliographical references and index.
Summary: An illustrated guide to seeds. Covers the internal structure and external architecture of seeds and follows the metamorphoses in their development from pollination to eventual germination. Includes photographs and electron microscopic images of seeds and their parent plants.
ISBN-13: 978-1-55407-221-7
ISBN-10: 1-55407-221-2
1. Seeds.   I. Stuppy, Wolfgang. II. Papadakis, Alexandra.   III. Title.
581.4/67 dc22   QK661.K477 2006

Library and Archives Canada Cataloguing in Publication

Kesseler, Rob
Seeds : time capsules of life / Rob Kesseler & Wolfgang Stuppy ; edited by Alexandra Papadakis.
Includes bibliographical references and index.
ISBN-13: 978-1-55407-221-7
ISBN-10: 1-55407-221-2
1. Seeds.   I. Stuppy, Wolfgang II. Papadakis, Alexandra III. Title.
QK661.K47 2006   581.4'67   C2006-900756-X

Published in the United States by
Firefly Books (U.S.) Inc.
P.O. Box 1338, Ellicott Station
Buffalo, New York 14205

Published in Canada by
Firefly Books Ltd.
66 Leek Crescent
Richmond Hill, Ontario L4B 1H1

Pulished in Great Britain by
Papadakis Publisher
16 Grosvenor Place
London, SW1X 7HH, UK

Published in association with the Royal Botanic Gardens, Kew

Printed and bound in Singapore

who planted the seed that gave rise to this book
Wolfgang Stuppy, Kew & Wakehurst Place, February 2006

In memory of George Busby (1926-2005)
artist, mentor, and friend who set me on the road
Rob Kesseler, London, February 2006

Acknowledgments

Creating the present book took not only two authors, a brilliant designer like Alexandra Papadakis, and a visionary publisher such as Andreas Papadakis, but also many colleagues, friends and partners who contributed in many different ways. We wish to thank Professor Sir Peter Crane, Director of the Royal Botanic Gardens, Kew, for his encouragement and support throughout the project. We thank the staff of the Royal Botanic Gardens, Kew, particularly the Micromorphology Section, all members of the Seed Conservation Department (SCD), and the many partners of the Millennium Seed Bank Project all over the world who have contributed towards the outstanding collection that provided the basis for this book. The Millennium Seed Bank Project is funded by the U.K. Millennium Commission and the Wellcome Trust. The Royal Botanic Gardens, Kew is partially funded by the U.K. Department of Environment, Food and Rural Affairs.

Picture Credits

Pages 23 and 38 Andrew McRobb, Royal Botanic Gardens, Kew; page 30, Prof. Dr. Thomas Stützel, Botanischer Garten, Ruhr-Universität Bochum, Germany; page 53, Prof. Dr. Lutz Thilo Wasserthal, Institut für Zoologie, Friedrich-Alexander-Universität, Erlangen, Germany, first published in *Botanica Acta* (Wasserthal 1997); page 135, Elly Vaes, Royal Botanic Gardens, Kew; page 183, Weald and Downland Museum, Sussex, photo Marc Hill; page 187, Apex News & Pictures, www.apexnewspix.com
Colour illustrations by Elly Vaes

We gratefully acknowledge the granting of permission to use these images. Every reasonable attempt has been made to identify and contact copyright holders. Any errors or omissions are inadvertent and will be corrected in subsequent editions.

cover illustrations: (front) *Spergularia media* (Caryophyllaceae) - greater sea-spurrey; collected in Belgium – seed with peripheral wing assisting wind-dispersal; diameter 1.5mm
(back) *Ornithogalum dubium* (Hyacinthaceae) – yellow star-of-Bethlehem; collected in South Africa – seed displaying an intricate surface pattern: the cells of the seed coat are interlocked like the pieces of a jigsaw puzzle; seed 1.1mm
page 1: *Castilleja exserta* ssp. *latifolia* (Orobanchaceae) – purple owl's clover; collected in California, USA – seed; 1.9mm long
page 2: *Ariocarpus retusus* (Cactaceae) – living rock cactus – flowering specimen; slow-growing, spineless and looking more like an agave; one of Mexico's most charismatic cacti that is much sought after by collectors
page 3: *Ariocarpus retusus* (Cactaceae) – seed; 1.5mm long

CONTENTS

Preface   8
HRH The Prince of Wales

Foreword   10
Professor Sir Peter Crane

Seeds: Time Capsules of Life   14

What is a Seed?   18

Seed Evolution   23

Naked Seeds   33

Flower Power Revolution   43

The Dispersal of Fruits and Seeds   90

Travellers in Space and Time   162

The Millennium Seed Bank   172

An Architectural Blueprint   176

Phytopia   190

APPENDICES   256

GLOSSARY   258

BIBLIOGRAPHY   261

INDEX OF PLANTS ILLUSTRATED   262

ACKNOWLEDGMENTS   264

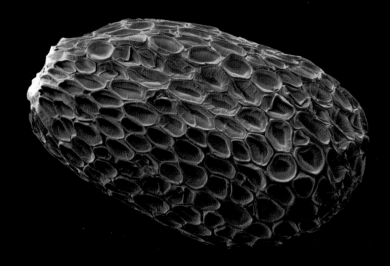

above: *Erica cinerea* (Ericaceae) – bell heather; collected in the UK – seed; 0.7mm long

Stellaria holostea (Caryophyllaceae) – greater stitchwort; collected in the UK – seed; 2.2mm long

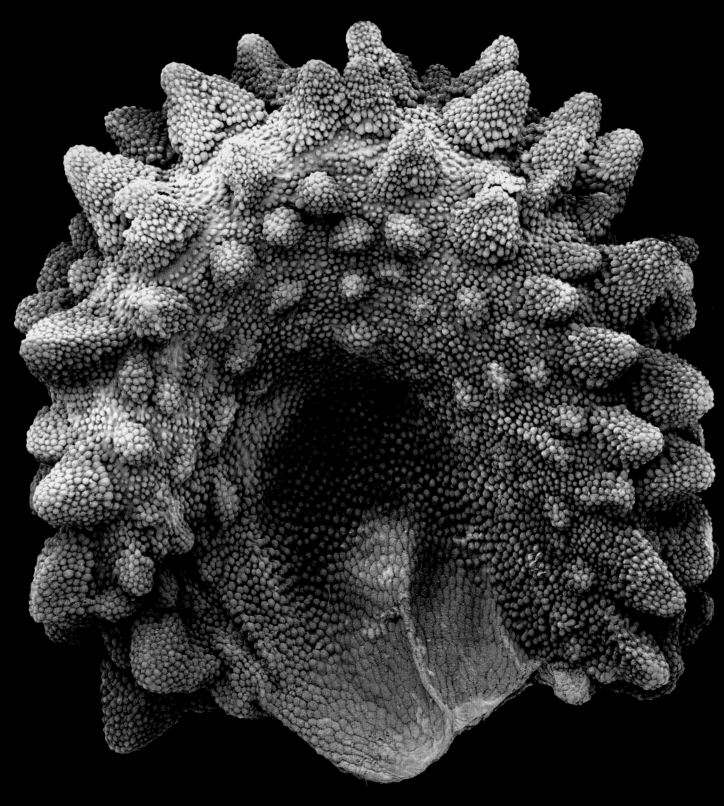

Thamnosma africanum (Rutaceae); collected in South Africa – seed; the reasons for the bizarre surface sculpturing are unknown; 2mm long

CLARENCE HOUSE

Anyone who has observed a snowflake through a magnifying glass knows that some of Nature's most exquisite handiwork is on a miniature scale. By combining the artistic eye with scientific observation at the microscopic level, this book reveals Nature's genius in the protection and adornment of seeds: those most vital of structures that secure the next generation of plants and, indeed, on which our whole civilization ultimately depends…

Seeds enabled our ancestors to make the transition from hunter-gatherers to settled agriculturalists capable of embarking on the journey of civilization. Seeds, as major food crops, still provide most of humanity with our most basic source of food. But seeds are also the means by which plants move through space and through time. Their extraordinary ability to lie dormant for tens, or even hundreds, of years provides us with a way to insure against the extinction of plants in the wild. All around us plant species are under threat as their habitats are lost or degraded through our unsustainable stewardship of the planet. Saving seeds for the future is an immensely important way of guaranteeing the future of mankind which, at the moment, is threatening with extinction one-quarter of the world's remaining 270,000 plant species by 2050.

I am delighted that this book is a collaboration involving Kew's Millennium Seed Bank Project, with which I have had a close association since its inception. Conservation of plant species for the long-term is the passion and purpose of the dedicated scientists who work at Kew. The beauty and knowledge captured by Rob Kesseler and Wolfgang Stuppy in this volume allow us to reflect on the glory as well as the utility of seeds, and the importance of conserving them – and the plants of which they are part - in all their infinite variety. After all, the consistent occurrence of five-fold (pentagonal) symmetry in wild flowers and seeds is a reminder of the intimate connection between the golden ratio, or golden mean, and the power of life and harmony.

Joy in looking and the sense of wonder in comprehension were, to Einstein, Nature's most beautiful gift. As we marvel at the exquisite beauty of the humble seed - who can disagree?

Saxifraga umbrosa (Saxifragaceae) – London pride, endemic to the western and central Pyrenees – seed; 0.6mm long

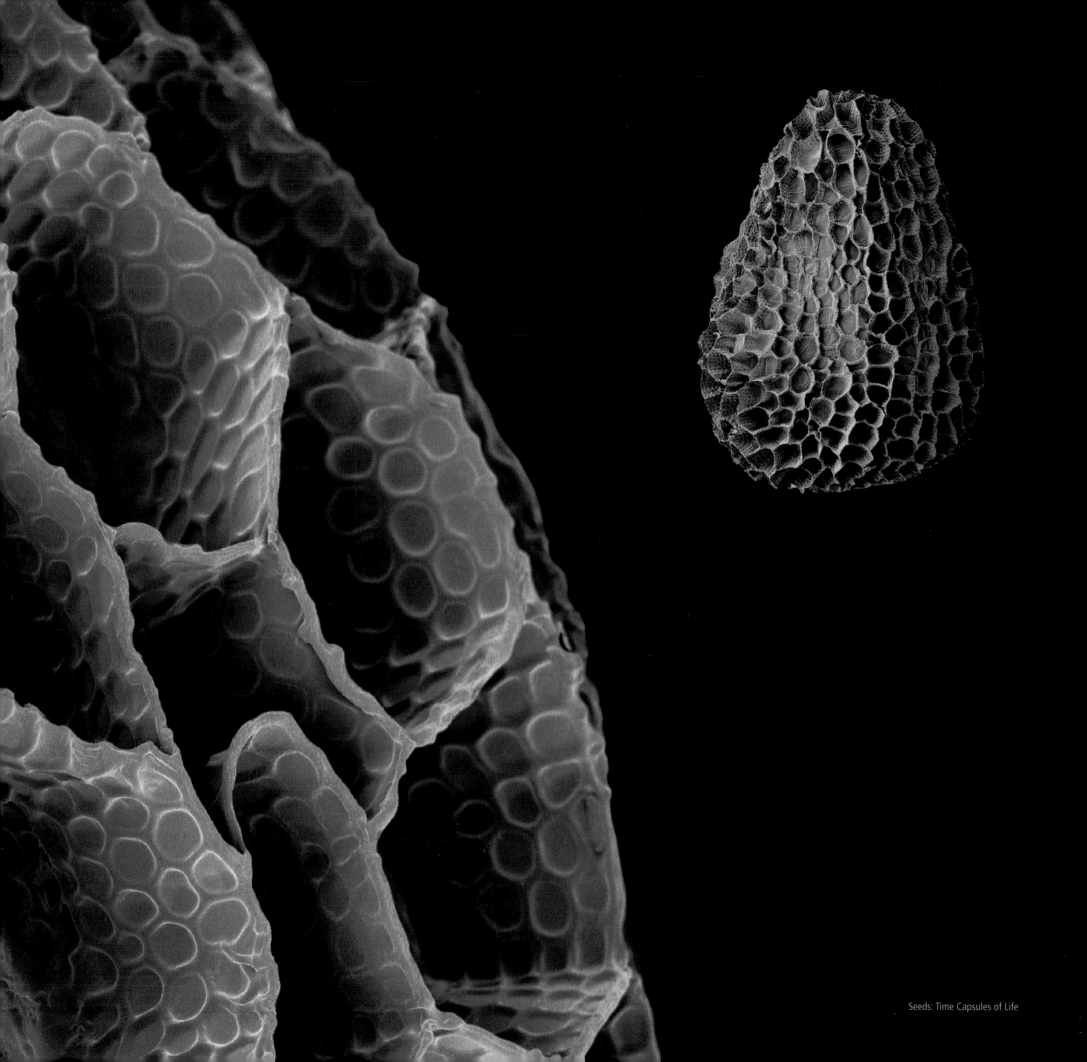

Seeds: Time Capsules of Life

The history of plant life on land extends back almost 600 million years, and across that great span of time the origin of the seed stands out as one of the key innovations that allowed plants to expand and diversify across the Earth's surface. Provisioned with food reserves, and with the ability to lie dormant before producing the next generation, seeds endowed plants with the capacity to survive temporary hardships and created new opportunities for plants to move from place to place. Every seed is a time capsule that adds resilience and mobility to the life cycles of plants: two properties that have had profound, long-term effects on the evolution of life on this planet.

In this marvellous book Rob Kesseler and Wolfgang Stuppy provide a lively account of seeds and their significance, while at the same time illuminating their beauty and diversity through a series of stunning images. The intricacy of these exquisite natural forms, often expressed at a microscopic scale, highlights the extraordinary creativity of Nature, and at the same time it is a reminder of how little we truly understand about the structure, biology and ecology of even the most common plants.

Over the broad sweep of evolutionary time, the resilience of seeds, that same feature that helps plants survive the harshest cold, the longest droughts and the darkness of polar winters, has imparted a degree of inbuilt extinction resistance. In natural seed banks, seeds can survive for many decades. In the ideal conditions of man-made seed banks this can be extended to centuries. The Millennium Seed Bank, and other seed banks around the world, already provide insurance against human-induced extinction for tens of thousands of different kinds of plants. It is to be hoped that these efforts will be extended and accelerated so that the creativity of millions of years of plant evolution is not irreplaceably lost during our own short tenure on this planet.

Given the urgency of current environmental concerns, engaging the public in the conservation of plants has to be a priority for any scientist interested in the variety of plant life. It is therefore central to the work of the Millennium Seed Bank and the Royal Botanic Gardens, Kew. We need to build public understanding of the importance of plants in all our lives. The creative partnership between artist and scientist in this book stimulates us to want to know more about seeds and how they work. I hope that it also encourages us to see the glory of Nature and to do what we can to ensure its survival into the future.

Professor Sir Peter Crane

Digitalis purpurea (Plantaginaceae) – purple foxglove; collected in the United Kingdom – seed (1.3mm long) and detail of seed coat

TIME CAPSULES OF LIFE

ROB KESSELER & WOLFGANG STUPPY

Paulownia tomentosa (Paulowniaceae) – princess tree; native to China – seed with a lobed peripheral wing to assist wind dispersal; 4.4mm long

There is a magical appeal, rooted in childhood, in watching seeds develop: the acorn in its neatly fitting cup or the polished, rich brown surface of the horse chestnut as it emerges from its spiny shell. These sensuous forms draw us closer to nature: temporal touchstones rolled between fingers, stuffed into pockets or left to slowly shrivel on windowsills. Then there is the poppy, with its flame red petals that quickly fall as the fruit ripens into its familiar capsule, the crop of seeds trapped inside, rattling like miniature maracas until the cap lifts and they are eventually dispersed.

Holding a small seed in one's hand it is sometimes difficult to comprehend that given the right conditions a complex and beautiful plant will emerge from it. Seeds are the beginning and end of the life cycle of plants, carriers of the genetic codes that will ensure successful propagation and continuation of the species. Their resilience is renowned: seeds taken from dried herbarium samples have been successfully germinated over two hundred years after they were collected. Their diversity of form and scale is as extensive as the plants from which they derive, from the giant *coco de mer* weighing up to twenty kilos to the almost dust-like seeds of the orchid family where one gram can contain more than 2 million seeds.

Until the seventeenth century the study of plants had largely been for medicinal or horticultural purposes, but taking advantage of the new compound microscope developed by chemist and physicist Robert Hooke, pioneering botanists such as Nehemiah Grew and John Ray were among the first to describe the structure and reproductive mechanisms of seeds. Fuelled with this new knowledge, a new breed of explorers and plant hunters were bringing back to Europe exotic flowers and plants to be cultivated by a growing number of botanists and plantsmen. This fuelled a competitive passion for growing flowering plants and subsequently for the gardens in which to display them, leading to a demand for ever more exotic varieties to fill the burgeoning hothouses and gardens of the nobility.

This growing passion laid the foundation for a more systematic approach to the collection and scientific study of plants with the creation of Botanic Gardens. In addition to living plants that miraculously survived the trials of being transported thousands of miles across land and sea, increasingly the collecting and trading of seeds became more commonplace. Today this has evolved into a multimillion pound industry to satisfy the demands of a highly educated population of garden enthusiasts. But more importantly, as environmental concerns have grown and the importance of the preservation of plant habitats for bio-diversity has been recognised, a network of highly trained seed collectors with local knowledge of endangered species has emerged. Their precious harvest is distributed among the many centres for botanical research around the world. In recognition of the urgent need for a concerted approach, the Royal Botanic Gardens, Kew, created the Millennium Seed Bank at Wakehurst Place in Sussex in 2000. The Millennium Seed Bank Project has set itself the daunting but vital task of collecting and conserving by 2010 over 24,000 species – of the world's seed-bearing flora.

In the eighteenth century, artists and scientists worked closely together to examine and portray the many complexities of life. In a revival of this collaborative spirit this book reunites the worlds of botanical science and art to reveal and celebrate the astounding diversity and complexity of seeds. As we worked together we marvelled over the specimens in front of us and through our collaboration we hope to show you things you may have seen but never had the opportunity to examine in minute detail. In the natural world seeds are dispersed on the wind, carried on the backs of animals or eaten by birds and other animals to be deposited far from the original plant. They are dispersed by humans too – as food transported across vast distances, as decorative items of jewellery, or accidentally when stuck to clothing. Through this book we hope to extend the strategy of dispersal to a new audiences.

WHAT IS A SEED?

WOLFGANG STUPPY

Lamourouxia viscosa (Orobanchaceae) – collected in Mexico – balloon seed displaying the most extreme form of the typical honeycomb pattern found in wind-dispersed seeds; both the outer and inner tangential walls of the cells of the seed coat dissolve to leave just the loose, voluminous honeycomb cage produced by the thickened radial walls; 1.2mm long

Seeds are time capsules, vessels travelling through time and space. In the right place at the right time each seed gives rise to a new plant. They are the most sophisticated means of propagation created by the evolution of plants on our planet and the most complex structure a plant produces in its life. Seeds have two principal functions: reproduction and dispersal. For most plants, the seed is the only phase in its life when it can travel. Each individual seed carries the potential of the whole plant and even of the species. Tiny herbs and giant trees both grow from seed. To maximise their success and to fill every available niche, plants have developed an incredible range of seed sizes, shapes and colours. For example, the largest seed in the world (in fact a single-seeded fruit), the Seychelles nut, *coco de mer* or double coconut, can be 50cm in length, nearly a metre in circumference and weighs up to twenty kilos. The smallest seeds are found in orchids and can be as small as 0.11mm long and weigh less than $0.5\mu g$ (= $0.0000005g$), which means that one gram contains more than two million seeds.

This amazing diversity, of which the tiniest examples are often of breathtaking beauty and exquisite sophistication, is largely the result of the pursuit of different strategies of dispersal. For a plant it is very important to find a way to disperse its seeds. This task can be accomplished either by the plant itself through exploding fruits that eject the seeds (the fruits of some legumes catapult their seeds as far as 60m) or by developing a range of astonishing adaptations that allow the seed to use wind, water or animals as transport vehicles. By enabling its offspring to travel away from the mother plant, even if for a short distance only, a species can conquer new territories, increase its numbers and alleviate competition between siblings and parents.

Imagine that children are being sent away to start a new life somewhere else, perhaps far away. What would they be given to make sure they stand a good chance of survival? They would certainly need something to eat, a packed lunch perhaps. They would probably also be given some kind of protection against the elements and, of course, against predators in search of the tasty morsel they are carrying. This is precisely the strategy of plants. A seed consists of three basic components: the offspring in the form of a small plant called the *embryo*; the energy-rich, nutritive tissue surrounding it, called the *endosperm*; and a protective layer around the outside, usually the seed coat.

It is this nutritive tissue, the food reserve of seeds, which makes them so precious, so indispensable for most human societies. It is no exaggeration that our entire civilisation is built on seeds. Think of cereals such as rice, wheat, maize, barley, rye, oat and millet; and pulses such as beans, peas and lentils. They are the main source of nourishment for billions of people worldwide and they are all seeds. Rice alone is the staple of half the people on

Cleome gynandra (Capparaceae) – cat whiskers; collected in Burkina Faso – shed from capsules, the seeds display no obvious adaptations to a particular mode of dispersal although their thick discoid shape and bizarre sculpturing may assist dispersal by wind and rainwater; 1.45mm in diameter.

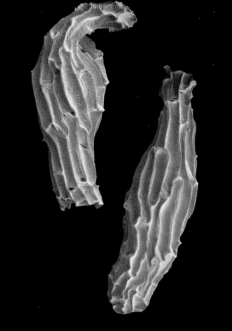

opposite: *Heracleum sphondylium* (Apiaceae) – hogweed; common in all northern temperate regions – close-up of dried fruit

above: *Ophrys sphegodes* (Orchidaceae) – early spider orchid; collected in the UK – seeds with loose, bag-like seed coat displaying the typical honeycomb pattern of wind-dispersed balloon seeds; 0.6mm long

below: *Lodoicea maldivica* (Arecaceae) – Seychelles nut; native to the Seychelles – the single-seeded nuts of this palm tree take 7-10 years to mature and contain the world's largest seed

Earth. Then there are the pleasures in life to which seeds contribute like the nuts we nibble, the beer we drink, or the coffee we crave in the mornings. Seeds provide many of the spices used in cooking: pepper, nutmeg, cumin, caraway, fennel and mustard, to name but a few. Seeds also yield precious raw materials: they provide valuable oils, like the linseed oil used in varnishes and paints; the rapeseed oil that serves as fuel; and castor oil, an excellent lubricant for jet engines and heavy machinery. Another raw material of great economic importance is cotton, which consists of the hairs shaved off the seeds of the cotton plant.

In addition to their immense usefulness, seeds can be extraordinarily beautiful. This book will change for ever the perceptions of those who have never appreciated their beauty.

Seed evolution

It is easy to comprehend how a seed is put together but few are aware of the complexity of seeds. Plants needed tens of millions of years to develop this most sophisticated means of sexual reproduction but it is a great success story. The evolution of seed plants from their primitive fern-like ancestors is a key chapter in the history of life on Earth. Understanding how and why seeds evolved, and how they have changed the face of our planet, makes the journey into the microscopic world of seeds both illuminating and exciting.

Where do seeds come from, why do they exist, and what makes them one of the greatest achievements of plant evolution? In order to answer these questions we have to take a brief look at the evolution of the life cycles of plants, and, most importantly, their methods of sexual reproduction. Primitive land plants such as mosses (including liverworts and hornworts), clubmosses, horsetails and ferns, the so-called *cryptogams*, neither have flowers[1] nor do they have the ability to produce seeds. They reproduce through spores.

Seeds and spores could hardly appear more different; for a long time it was believed that they have nothing in common. But in 1851, the self-taught German botanist Wilhelm Hofmeister (1822-77) made a famous discovery: he demonstrated that the *alternation of generations* in seed plants follows the same principle as in mosses and ferns and thus proved their evolutionary link. To grasp the significance of Hofmeister's ingenious discovery and to understand how seeds evolved, it is necessary to take a closer look at the private life of seed plants.

It's all about sex

Just as human life starts with the union of a sperm from the father and an egg cell from the mother, with each parent contributing one set of chromosomes, a new plant is created when a male sperm and a female egg cell meet. In all living beings, the chromosomes

Seeds: Time Capsules of Life

opposite: *Dryopteris filix-mas* (Dryopteridaceae) – male fern; collected in the UK – close-up of sorus with dehisced sporangia around the shrivelled indusium; sporangium diameter: c.0.2mm

above: *Dryopteris filix-mas* (Dryopteridaceae) – male fern; collected in the UK – content of one sporangium surrounded by loose spores from neighbouring sporangia; cluster diameter: 0.15mm

below: *Dryopteris filix-mas* (Dryopteridaceae) – male fern; collected in the UK – underside of a fertile frond showing the brown sori (clusters of sporangia)

contain the genes that determine all the characteristics of the new organism. By mixing the chromosomes – and thus the genetic traits – of two different individuals, a new organism with a different and perhaps better combination of characteristics is created. It is sex that provides the "raw material" for evolution.

Meiosis and the legend of the rice grain on the King's chessboard

The key to sexual reproduction is a sophisticated way of cell division called *reduction division* or *meiosis*. Meiosis is a universal process in all sexually reproducing organisms, both plants and animals. It reduces the number of chromosomes in the *gametes* (sperm and egg cells) to half, either directly (in animals) or indirectly (in plants via the gametophytes). Without it, the number of chromosomes would double in each new generation, like the rice grain in the famous mathematical folktale from India: a long time ago, a wise man performed a service for the King of Deccan. The King insisted on rewarding him and, after some hesitation, the humble man asked for a single grain of rice for the first square on the King's chessboard, two for the second square on the second day, four for the third square on the third day, and so on. The number of rice grains would be doubled every day until every square of the chessboard was filled. The king, who had never heard of exponential growth, agreed, foolishly as he would soon discover. He owed the man 18 million billion rice grains, or, more precisely, 18 446 744 073 709 551 616 (2 to the 64th power) – more rice than could be grown on the entire surface of the Earth including the seven seas.

The same would happen to the number of chromosomes in subsequent generations if meiosis did not precede each act of sexual reproduction: for sexual reproduction to function, gametes (egg cells and sperm cells) must be *haploid* (contain only one set of parental chromosomes) so that they produce a new *diploid* organism, which contains no more than two sets of chromosomes.

Alternating generations

Some plants are capable of vegetative propagation (by producing suckers, like strawberries, for example), which does not create a new kind of individual but simply clones of the mother plant, but most plants reproduce sexually. Just like in humans and most animals, a sperm has to fertilise an egg cell to produce the next generation. For green algae – one group of which was the ancestor of our land plants – fertilization was never a great problem. Their aquatic lifestyle allowed them to release their sperm cells into the water, where they could swim freely to find an egg cell; a simple and effective, albeit hugely wasteful method of fertilization.

When plants left the water and began to conquer the land, most probably at the end of the Ordovician or at the beginning of the Silurian (about 445 million years ago), they had to adapt to the drier conditions of life in the open air. Having mobile sperm that require water to swim across to an egg cell to be fertilized was a real disadvantage when living on land. Today's spore-producing land plants, such as mosses, clubmosses, horsetails and ferns, have yet to solve this problem. Their distribution is limited by the requirements of the egg- and sperm-producing phase of their life cycle. This is why they are usually found growing in humid environments or in areas where wet periods are common in otherwise dry conditions (e.g. xeric ferns like *Cheilanthes*).

Despite this handicap, these spore-producing plants possess an effective method of sexual reproduction involving a cycle called *alternation of generations*,[2] something not found in animals. Alternation of generations means that their life cycle involves two generations, a diploid generation (with two sets of chromosomes, one set from each parent) and a haploid generation (with only half the number of chromosomes), each of which can only give rise to the other. And this is where the difference between seeds and spores becomes obvious. Seeds germinate to produce a new diploid generation, whereas spores produce a haploid generation. This may sound complicated but a comparison of the life cycles of the different types of land plants will clarify this fundamental difference between seeds and spores starting with the most primitive land plants, which are mosses.

The life cycle of mosses

Moss spores give rise to the haploid generation, the small familiar moss plants. When they reach maturity, they produce both male organs (*antheridia*) that release mobile sperm, and female organs (*archegonia*) containing egg cells. Sperm and egg cells are generally called *gametes*, which is why the small moss plant that produces them is also referred to as the *gametophytic generation* or *gametophyte* (literally *gamete plant*).

A romantic swim

In the presence of water (rain, dew, spray from a river or waterfall), the sperm cells are released from the antheridia on the male plants and swim across to the egg cells waiting for them in the archegonia on another plant. Like the gametophytes which produced them, the sperm and the egg cell are both haploid and contain only one set of chromosomes. After fertilization the egg cell contains two sets of chromosomes (i.e., it becomes diploid) and is called a *zygote*. The zygote stays on the mother plant, which provides it with water and nutrients, and as it develops it will produce a tiny moss embryo. This embryo grows into

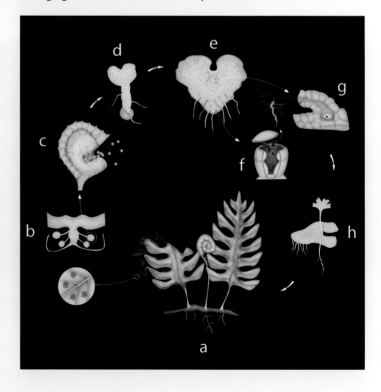

Fern life cycle – clockwise (a) sporophyte; (b) sporangia on the underside of the fertile fern frond covered by the indusium; (c) mature sporangium releasing spores; (d) germinating spore; (e) mature gametophyte (prothallus) producing both antheridia (male sexual organs) and archegonia (female sexual organs) on the underside; (f) mature antheridium releasing motile sperm, which swim to the nearest mature archegonium; (g) archegonium containing a single egg cell; (h) young sporophyte emerging from the underside of the prothallus

the familiar moss capsule that represents the diploid generation of the moss life cycle. Since new haploid spores are produced inside the moss capsule, it is also called the *sporophytic generation* or simply *sporophyte* (literally *spore plant*). When the capsule is ripe, it opens at the top and releases the spores to be blown away by the wind or washed away by water. In a suitable location, the spores grow into a new moss plant (gametophyte) and the reproductive cycle starts anew.

The life cycle of ferns

In principle at least, ferns have the same sex life as mosses. In the right place and with enough moisture, a fern spore will produce the gametophytic generation. However, the gametophyte of a fern does not look like a fern at all. It is a small, green, flat lobe very similar to the gametophyte of some liverwort. This haploid plantlet, also called a *prothallus* or *prothallium* (literally *pre-shoot*), usually produces its sex organs (antheridia and archegonia) on its underside to prevent them from drying out. Although fertilization in ferns follows the same pattern as in mosses, what happens afterwards puts every moss to shame: the zygote of a fern does not just produce a simple small capsule on a stalk that needs the support of the mother plant. Instead, it develops into a totally independent, impressive and often very beautiful plant that can grow for many years, sometimes becoming the size of a tree. This means that in the life cycle of ferns, the gametophytic generation remains relatively poorly developed whereas the sporophytic generation is strongly enhanced.[3] Hence, every fern plant we see is a sporophyte. There are often regular rows of small, slightly raised, brown spots on the underside of fern fronds. These brown spots, called *sori* (singular *sorus*), are groups of spore containers (*sporangia*; singular *sporangium*) covered by a protective umbrella, the *indusium*. Like moss spores, fern spores are minute and float easily in the air, sometimes travelling for miles on a light breeze.

What does the sporophyte have that the gametophyte does not?

A life cycle favouring the sporophyte has a clear evolutionary advantage that lies in the genetics of the plant: the sporophyte is diploid and thus has two copies of each gene. This means that if one gene malfunctions owing to a mutation, the plant has a second copy of the same gene, which acts as a back-up and compensates for the damage. Genetic mutations are therefore less likely to have a negative impact on the vitality of a sporophyte than on the vitality of a gametophyte. Moreover, two slightly different copies of the same gene allow the sporophyte to respond more flexibly to changes in the environment resulting in better fitness than in the gametophyte.

Ampelopteris prolifera (Thelypteridaceae) – fern; native to the Old World tropics – young sporophyte emerging from the underside of the dark green liverwort-like prothallus

Seed Evolution

When males are *micro* and females are *mega*

While most mosses are *monoecious*, meaning they bear antheridia and archegonia on the same gametophyte, some are *dioecious*, meaning they produce sperm and egg cells on different individuals. Most ferns prefer the former. Only a very small group of ferns, the *water-ferns* to which the water clover (*Marsilea, Regnellidium*), the pillwort (*Pilularia*), the duckweed fern (*Azolla*) and the floating fern (*Salvinia*) belong, produce separate male and female gametophytes. The sporophyte of these ferns must therefore produce two kinds of spores: male and female. Apart from their gender, male and female spores also differ in size, a condition called *heterospory*. Because of this difference in size, the male spores are called *microspores* and the female spores *megaspores*. Microspores give rise to male *microgametophytes* and megaspores to female *megagametophytes*. The containers in which these two types of spore are produced on the sporophyte are *microsporangia* and *megasporangia*. The leaves of the sporophyte which bear these micro- and megasporangia are called *micro-* and *megasporophylls*, respectively. This avalanche of technical terms may be daunting but it makes comparisons of the different life cycles of plants much easier. For now, it is enough to remember that *micro* means *male* and *mega* means *female*.

In present-day water-ferns, heterospory almost certainly represents an adaptation to the aquatic lifestyle to which they reverted. Nevertheless, it was the preferred condition among the ancestors of our seed plants. In fact, as will soon become clear, heterospory played an important role in the evolution of seeds.

How seeds evolved

The first seed plants or *spermatophytes* appeared some 360 million years ago, towards the end of the Devonian. These early seed plants combined seeds with fern-like foliage and were thought to be intermediate between ferns and modern seed plants, hence the name *seed ferns* or *pteridosperms*. However, it is now known that ferns themselves were not the direct ancestors of seed plants. This role falls to an extinct, rather obscure side branch of heterosporous fern-like plants. The exact events and transitional stages that led from heterosporous fern-like ancestors to the earliest seed plants are uncertain. What is certain is that the two crucial steps towards the seed habit were the evolution of the *ovule* and the *pollen grain*.

From megasporangia to ovules

The transition from spore-bearing fern-like plants to seed-bearing plants was marked by significant changes in the megasporangium and the megagametophyte. Unlike their heterosporous fern-like ancestors, which dispersed their spores freely on wind and water,

top: *Azolla filiculoides* (Azollaceae) – duckweed fern; found in all tropical and warm regions
centre: *Salvinia* sp. (Salviniaceae) – floating fern; found in all tropical and warm regions – a dense covering of water-repellent hairs on the upper leaf surface makes the leaves unwettable. The hairs consist of a stalk, which divides into four branches that reconnect at the tip to form a cage-like air-trap
above: *Marsilea quadrifolia* (Marsileaceae) – European waterclover; native to Eurasia – the leaves of this water fern are composed of four lobes resembling clover

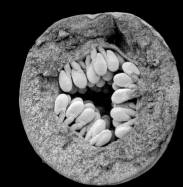

top: *Passiflora* cv. 'Lady Margaret' (Passifloraceae) – passionflower 'Lady Margaret' – immature ovules, their nucellus (bright red) surrounded by the developing outer and inner integument, which at this very early stage form two collars at the base rather than a complete envelope; each ovule 0.1mm in diameter

centre: *Passiflora caerulea* (Passifloraceae) – blue passionflower; native to South America – cross-section of ovary with mature ovules; cross-section diameter: 3.4mm

above: *Passiflora* cv. 'Lady Margaret' (Passifloraceae) – ornamental passionflower in cultivation at Kew

seed plants retained their megaspores within the megasporangia. The megasporangia themselves remained attached to the sporophyte and each produced only a single viable haploid megaspore. From this single megaspore, the female gametophyte developed within the confines of the megasporangium. Another very important evolutionary change was that the sporophyte of seed plants covered its megasporangia with a protective layer, called the *integument*. The integument may have evolved from the fused primeval branches (the *telomes*) that surrounded the megasporangium and provided the coat of the mature seed. Through the retention and nurture of the megagametophyte on the mother plant and the addition of a protective integument, the megasporangium had evolved into a new improved organ, the *ovule*. Although very small and remaining within the megasporangium (now called the *nucellus*), the megagametophyte continued to produce archegonia-bearing egg cells, which needed to be fertilised. But with the megagametophyte developing on the sporophyte inside the nucellus rather than freely on the ground, the transfer of sperm became more difficult. This problem was solved by the evolution of *pollen* (Latin for *fine flour*).

From microspores to pollen grains

Since the megasporangium remained attached to the sporophyte and the megagametophyte developed inside the ovule, fertilization in seed plants had to take place directly on the sporophyte and the microspores of the early seed plants had to adapt to new conditions. As the megagametophytes were now up in the air and still attached to the sporophyte, the absence of water was more likely to be a limiting factor for fertilization. A new method of transfer, independent of water, was needed to allow the sperm to reach the egg cell.

The solution came with the evolution of pollen. Pollen grains are simply tiny (micro-) spores that are able to germinate on or near the megasporangium to produce a very small and greatly simplified microgametophyte. The pollen-producing microsporangia of seed plants are called *pollen sacs*. The pollen grains originate from fertile diploid tissue inside the pollen sac, the *archespore*. At an early stage in the development of the pollen sac, the diploid cells of the archespore turn into *microspore mother cells* (or *pollen mother cells*). Each of these pollen mother cells undergoes a reduction division (meiosis) and produces four haploid microspores, the pollen grains. At the time of pollination, the pollen sacs open and release the pollen grains into the environment.

Primeval pollination

The earliest known seed plants – such as *Moresnetia zalesskyi* (named after the town of Moresnet in Belgium) and *Elkinsia polymorpha* (named after Elkins, a small town in Wes

Virginia, U.S.A), which date from the Devonian (417-354 million years ago), had a peculiar method of pollination and rather strange looking ovules.

Their primitive *pre-pollen* consisted of slightly more advanced wind-dispersed microspores, destined to land and germinate on the ovules where they almost certainly released motile sperm. The ovules of these early seed plants were also very primitive. In these *pre-ovules*, which measured about 1-2mm in diameter and 3-7mm in length, the integument did not yet form a complete envelope around the megasporangium (=nucellus) but consisted of several separate spreading lobes that surrounded the nucellus like a cup, leaving its top visible. At its exposed apex the nucellus of pre-ovules had a funnel-shaped opening, the *lagenostome*. The function of the lagenostome was to collect pollen from the air. Some believe that pollen grains landed in the funnel from where they were passed directly into a pollen chamber below. But it is more probable that both the pollination chamber and the lagenostome were filled with liquid and the pollen was captured via a pollination drop with a meniscus that arched over the opening of the lagenostome. The pollination drop was reabsorbed and the captured pollen sucked into the pollen chamber above the area where the megagametophyte produced its archegonia. At the bottom was the *central column*, a protrusion formed by nucellar tissue. Once sufficient pollen had dropped through the lagenostome or the pollination drop had been reabsorbed, the pollen chamber was sealed off against the outside in a most remarkable way. The megagametophyte grew a "tent-pole" that ruptured the megasporangium wall covering the floor of the pollen chamber, and pushed the central column upwards to plug the opening of the lagenostome. Now in contact with the megagametophyte and its archegonia, the pre-pollen grains germinated, presumably to release motile male gametes very similar to the ones produced by their fern-like ancestors. The watery liquid in which they swam in the pollen chamber (probably the remains of the pollination drop or some kind of similar secretion) was produced by the ovule.

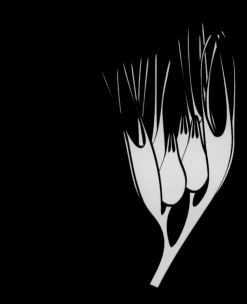

Sperm travelling by tube

It is not clear what the microgametophytes of the earliest seed plants looked like. At the very latest in the Upper Carboniferous (about 300 million years ago), the pollen grains of seed plants are known to have germinated with cylindrical outgrowths, the *pollen tubes*. Pollen tubes were initially formed as *haustoria* that grew into the tissue of the nucellus to absorb nutrients with which to support the growth of the microgametophyte. This is still the prevailing condition in today's cycads (a group of ancient seed plants superficially resembling palms) and *Ginkgo*. Their male gametophytes develop into haustorial pollen tubes, which grow over a period of several months and penetrate the nucellus. In cycads

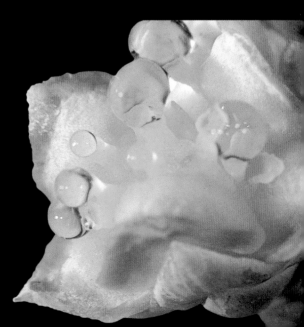

and *Ginkgo* the male gametes are still motile and two large swimming sperm are released into the pollen chamber from where they make their own way to the archegonia with the egg cells. The microgametophytes of *conifers* and other seed plants use the energy stored in the pollen grain to grow. Their male gametes lack flagellae and are unable to move, so they are transported down the pollen tube straight to the egg cell, independently of water.

The micropyle – gateway to the egg

During their evolution, the ovules of seed plants also experienced significant changes. Between the late Devonian and early Carboniferous, the integument of some spermatophytes (seed plants) began to form a single cohesive layer enclosing the entire nucellus. In order to allow pollen continued access to the lagenostome, the integument had an opening at the apex, the *micropyle*. Although today's seed plants have long lost the lagenostome, the micropyle is still a distinct feature of their ovules and remains the gateway to the egg cells. A pollen chamber can still be found underneath the micropyle of present-day *cycads* and conifers, where it exudes a sticky fluid, the *pollination drop*. This pollination drop captures pollen from the surrounding air. As the pollination drop is reabsorbed, it sucks the collected pollen grains through the micropyle into the pollen chamber where they germinate and finally release the male gametes from their pollen tubes.

To boldly go where no plant has gone before

The evolution of the pollen and the ovule were the most fundamental steps forward in the evolution of land plants. It won them independence from water for their sexual reproduction, giving them an enormous advantage over all other plants that had hitherto existed. In seed plants, the fertilised egg cell develops into a new sporophyte within the safety of the ovule. Unlike ferns, where the zygote has to grow into a new sporophyte immediately, once the embryo of seed plants has reached a certain size it often rests for a while inside the ovule until conditions for germination improve. This temporarily inactive embryo is equipped with a food reserve by the mother plant and protected by the integument (better known as the *seed coat*) against desiccation and physical damage. The seed had arrived and land plants were ready to boldly go where no plant had gone before, expanding into almost every corner of the planet, from the poles to the equator, wherever plant-life was possible.

Climate change, safer sex and the rise of seed plants

The significance of the evolution of the seed among plants can be compared to that of the

opposite above: **Archaeosperma arnoldii**, a pteridosperm from the late Devonian – reconstruction of a seed-bearing branch showing two pairs of seeds surrounded by cupules

opposite below: **Fitzroya cuppressoides** (Cupressaceae) – Patagonian cypress; native to Chile and Argentina – young female cone with mature ovules presenting a pollination drop at the micropyle

below: Primeval pollination – diagram illustrating one theory of how the first seed plants were pollinated: pre-ovules (shown in longitudinal section) were surrounded by an integument consisting of several separate spreading lobes. The nucellus had a funnel-shaped opening at its apex, the lagenostome. Most probably pollen was captured via a pollination drop whose meniscus arched over the opening of the lagenostome. The pollination drop was re-absorbed and the captured pollen sucked into the pollen chamber, which was then sealed by a protrusion formed by nucellar tissue. Inside the sealed pollen chamber, in close proximity to the megagametophyte, the pre-pollen grains germinated and presumably released motile sperm

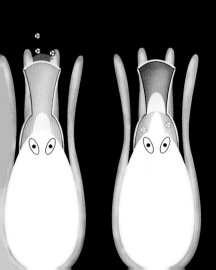

dependency on moist habitats, the egg enabled reptiles to become the first fully terrestrial vertebrates. In that sense, mosses and ferns are more like amphibians, relying on water for fertilization despite their terrestrial existence. Like all radically new innovations, it took some time for seed plants to dominate the flora of our planet. The first were humble creatures, small trees at best, which produced seeds at the tips of branches not in specialised structures such as the cones of the more advanced conifers and cycads. In the time period after the Devonian (417-354 million years ago) and the Carboniferous (354-290 million years ago), seed plants were still overshadowed by their giant spore-bearing contemporaries.

In the Palaeozoic era, in the geological time periods of the Devonian and Carboniferous, the Earth's climate was generally warmer and more humid than today making it ideal for spore-producing plants dependent on water for their sexual reproduction, and allowing them to grow to a gigantic size. During the Carboniferous, the Earth was covered by giant forests that thrived in the extensive swamps occupying large parts of our planet. These swamp forests consisted of tree-like heterosporous clubmosses and horsetails, tree-ferns and other plants extinct today. The most outstanding members of this long-lost flora were the tall scale-trees (up to 35m), such as *Lepidodendron* (Greek *lepis* = scale + *dendron* = tree) and seal-trees, such as *Sigillaria* (Latin: *sigillum* = seal), giant relatives of today's clubmosses and quillworts which dominated the prehistoric swamplands. It was mainly these heterosporous fern-like trees that provided the organic matter later converted into coal.

In the Permian (248-290 million years ago), most continents came together in a super-continent known as *Pangaea*. The formation of this enormous land mass triggered global cooling, creating a more extreme, arid environment, especially in the interior of Pangaea. Ecosystems were dramatically changed. Coal swamps mostly disappeared, as did 70% of land vertebrates and 85% of ocean species, in the most catastrophic mass-extinction in the Earth's history. This was the hour of the seed plants. The new climate conditions were far from ideal for spore-producing plants. Seeds afforded a much safer way of sexual reproduction, independent of water – an enormous advantage in the drier environment of the Permian.

During the late Carboniferous and following Permian, seed plants became large trees and soon displaced their cryptogamic contemporaries from nearly all habitats. Today, 97 per cent of all land plants belong to the *spermatophytes (seed plants)*. Seeds were simply too good an invention: obsolescent models such as scale-trees and seal-trees could not compete.

The competition never stops
In principle, evolution follows the rules of business: success can be maintained only by constant innovation that creates better products more efficiently. In the course of evolution

the advent of pollen and seeds triggered further adaptations towards ever more economic ways of sexual reproduction among the spermatophytes. Whilst the spores of the ancestors of seed plants – and those of mosses, horsetails, clubmosses and ferns – are able to find a suitable place for germination anywhere where there is sufficient moisture, seed plants require a much more precise deposition of their pollen. Many changes in the architecture of the reproductive organs of seed plants can be understood only as adaptations to improve the release, transfer and reception of pollen.

Naked seeds

The ovules of the early seed plants were borne "naked" on branches or along the margins of leaves (*sporophylls*), which is why they were given the name *gymnosperms* (literally "the ones with naked seeds"). This primitive arrangement of ovules can be observed today in the sago palm, a tree belonging to the most primitive extant genus of cycads known as *Cycas*. In female specimens, circles or clusters of ovule-bearing sporophylls alternate with normal leaves at the apex of the plant, astonishingly old-fashioned reproductive behaviour.

The seed-bearing organs of the gymnosperms

Already some of the earliest pteridosperms (including *Elkinsia* and *Moresnetia*) protected their ovules by partially surrounding them with loose cup-like structures, called *cupules*. These cupules were about 10mm long and 7-10mm wide, deeply lobed and contained between one and four ovules. Apart from their protective function, their extended lobes probably also assisted in the capture of pollen by causing eddies in the moving air.

Another method of protecting both ovules and pollen was to arrange the sporophylls in *cones*. Apart from members of the genus *Cycas*, all other cycads (and, strangely, all male *Cycas* plants), as well as conifers, crowd their pollen-producing leaves (*microsporophylls*) and ovule-bearing leaves (*megasporophylls*) together, but separated by gender, in male pollen cones and female seed cones. This tight arrangement of hardened sporophylls on specialised side shoots offers better protection of their precious pollen and ovules against animal predators like beetles. But when the time for pollination arrives, their ovules still have to be exposed to the environment in order to capture the pollen necessary for fertilization. Conifers and cycads are therefore still deemed to be naked-seeded gymnosperms. The ancient appearance of cycads is fascinating but not surprising considering the evolutionary age of the group. Fossils of cycads, including members of the most ancient-looking genus *Cycas*, can be found in sediments dating back to the earliest Permian (248-290 million years ago). During much of the following Mesozoic era (spanning the Triassic, Jurassic and Cretaceous), cycads were so abundant and diverse that this period is often referred to as the "age of cycads and dinosaurs". Since then cycads have declined and what is left today is a mere fraction of their Mesozoic diversity. They survive as living fossils, almost unchanged for more than 200 million years, with only some 290 species. Conifers appeared a few million years earlier than the cycads, probably in the late Carboniferous. During the Permian and Mesozoic they dominated many forest ecosystems from tropical to boreal climates. For reasons that will be revealed later, conifers suffered a similar decline to the cycads and survive with only 630 extant species.

The biggest cones in the world

In extant cycads, each female sporophyll bears two ovules near its base. The only exception is *Cycas,* whose primitive leaf-like megasporophylls carry between four and ten ovules, depending on the species. Cycad cones, especially the female ones, can be very large and heavy, as the two members of the genus *Lepidozamia* (Zamiaceae) from eastern Australia impressively demonstrate. *L. hopei* (Hope's cycad) can grow to 29m and is the tallest living cycad. Like all other cycads, it is *dioecious*, which means that it produces pollen cones and seed cones on separate plants. Its pollen cones can be up to 70cm long and the seed cones as long as 80cm, with a weight of up to 45kg. Probably the largest (up to 90cm long) and heaviest (more than 45kg) seed cones of all the cycads are produced by its close relative *L. peroffskyana,* also evocatively known as the pineapple zamia. Compared with the cycads, the longest conifer seed cone is rather small at 25 to 50cm; it belongs to the North American sugar pine (*Pinus lambertiana*, Pinaceae). Shorter in length, but much more massive are the seed cones of the Coulter pine (*P. coulteri*) from California. They can be up to 35cm long and weigh more than 4kg – rather dangerous when they drop off the tree.

The enigma of the female conifer cone

The pollen cones of conifers are usually relatively puny, rather soft and short lived. Their female counterparts are much larger and harder because of their woody scales, and can stay on the tree for several years after fertilization. Female conifer cones are – as one would expect – a little more complicated than the male cones. Like the male cycad cone, the male conifer cone is simply a short branch bearing microsporophylls with many (in cycads) or just two (in conifers) pollen sacs on the underside. A single individual scale of a female conifer cone is deemed to be a greatly reduced lateral branch. This sounds rather complicated and in fact the interpretation of the female cone of conifers has been (and for some still is) the subject of many arguments among botanists. After all, each scale of a conifer seed cone bears two ovules, just like the scale of a female cycad cone, so why are they not

opposite: *Encephalartos laevifolius* (Zamiaceae) – native to South Africa and Swaziland – female plant; seed cones c.20-30cm long and 15-20cm in diameter

below: *Pinus lambertiana* (Pinaceae) – sugar pine; cone collected in California – the seed cones of this pine species from Western North America are the longest conifer cones in the world; 25 to 50cm

the same? A comparison of modern conifers with fossils of their extinct ancestors demonstrated the complex nature of the seed cone and allayed the concerns of most botanists.

Weird and wonderful gymnosperms

Apart from the familiar cycads and conifers, there are other rather special gymnosperms. One of them is the unique ginkgo from China, also called the maidenhair tree because its fan-shaped leaves vaguely resemble the leaflets of a frond of the maidenhair fern (*Adiantum*). The ginkgo is a unique species with no living relatives and therefore classified in its own genus (*Ginkgo*), own family (Ginkgoaceae), own order (Ginkgoales), own class (Ginkgoopsida), and own division (Ginkgophyta). Petrified leaves of ancestral species have been found in sediments dating back 270 million years, to the Permian, which makes the ginkgo another example of a living gymnosperm fossil. During the middle Jurassic and Cretaceous period, about 175-65 million years ago, gingkos were widespread and several species occurred throughout the ancient continent of Laurasia (today's North America and Eurasia). Relatively recently (in geological terms), by the end of the Pliocene (5.3-1.8 million years ago), ginkgos disappeared from the fossil record everywhere. Only the modern species, *Ginkgo biloba*, survived in a small area in south-east China, where it has long been revered as a sacred tree by Buddhists and planted in temple gardens. Because of their resilience in the face of pollution, ginkgos are popular trees all over the world. Like cycads, they are dioecious and the male plants at least arrange their microsporangia in small cones. Female individuals carry their ovules in pairs at the tips of short stalks.

Other unusual and mostly unnoticed gymnosperms are members of the *Gnetales*, which include only three living genera, which do not resemble each other and each is placed in its own family: *Gnetum* (tropical broad-leaved vines or trees, deceptively looking like an angiosperm), *Ephedra* (green-twigged shrubs of the Western Hemisphere and Eurasian semi-deserts, best known as the source of ephedrine, a medicine for colds), and the weird and wonderful *Welwitschia mirabilis* from the deserts of south-west Africa, one of the most bizarre plants on our planet. *Welwitschia* was named after the Austrian botanist Friedrich Welwitsch (1806-1872), who found the first plants in 1860 in the Namib Desert in southern Angola. Its Latin species name *mirabilis* means wondrous; this needs no further explanation once one has seen the plant. Like cycads and conifers, *Welwitschia mirabilis* bears its seeds in conspicuous cones. The plant itself consists of a huge underground taproot reaching deep down to the water table. All that is visible above ground is a stout, cup-shaped stem with two strap-shaped leaves. Constantly growing at the base while dying back from the tips, these are the only leaves the plant produces in its life, which can last for 1,500 years.

opposite: *Cycas revoluta* (Cycadaceae) – sago palm; native to Japan – apex of female plant with an emerging cluster of ovule-bearing megasporophylls. Unlike male specimens of the genus, female *Cycas* plants do not arrange their sporophylls in cones but produce them at the apex alternating with normal leaves

below: *Cycas megacarpa* (Cycadaceae) – native to Queensland – close-up of megasporophylls with developing seeds

The sex life of gymnosperms

Despite great innovations such as pollen and the ovule, the sex life of gymnosperms still shows remarkable similarities with that of their spore-bearing ancestors, especially in the behaviour of the female gametophyte, which in the gymnosperms is a substantial structure that consists of thousands of cells. It develops from the haploid megaspore, which originates from the meiotic division of the megaspore mother cell.

The mother of a spore

During meiosis, one cell of the tissue inside the megasporangium, the diploid *megaspore mother cell* or *megasporocyte*, forms a linear tetrad (row of four) of haploid megaspores. Only one of the megaspores survives to divide many times, producing a large fleshy female gametophyte. This megagametophyte grows at the expense of the tissue of the megasporangium but remains enclosed within its remains. In the same way that the nucellus produces a haploid egg cell, the pollen sacs (microsporangia) produce haploid pollen grains through meiotic division of the pollen mother cells in the archespore. The only difference is that all four microspores arising from one pollen mother cell survive to form pollen grains. The germinated pollen grain then represents the reduced microgametophyte. As a result of its miniaturisation, the male gametophyte of the seed plants has long since ceased to produce its sperm in special antheridia but some eerie reminders of the past remain.

The biggest sperm in the world

Although seed plants have largely solved the sperm's transport problem, the most primitive spermatophytes living today (ginkgos and cycads) still produce mobile sperm cells that swim the last few millimetres through a watery liquid produced by the female ovule. They propel themselves forward with small tails, called *flagellae*, just like the sperm cells of their distant fern relatives. These "swarm spores" move and in fact look like small, unicellular animals and are therefore also referred to as *spermatozoids*. They are, in fact, rather like human sperm, but bigger and with many more tails. In *Zamia roezlii*, an interesting cycad from the coastal Choco region in Colombia, a single sperm cell is almost half a millimetre (0.4mm) in diameter (visible to the naked eye) and has 20,000 to 40,000 flagellae at the back. The synchronous pulsating beat of these flagellae propels the spermatozoids forward on their way towards the egg cell. With such gigantic spermatozoids the cycads hold the world record among living organisms.

In all seed plants, the male gametophyte produces just two male gametes (spermatozoids or sperm nuclei). Only the enigmatic Cuban cycad *Microcycas calocoma* releases some 12 to

opposite above: *Ephedra equisetina* (Ephedraceae) – bluestem joint fir; native to China and Japan – fruit consisting of two pairs of fleshy red bracts surrounding a single seed

opposite below: *Welwitschia mirabilis* (Welwitschiaceae) – tree tumbo; native to south-west Africa – seed cone

below: *Ginkgo biloba* (Ginkgoaceae) – ginkgo, maidenhair tree; native to China – mature seeds borne in pairs on a common stalk. The outside consists of the yellow fleshy sarcotesta, which contains butyric acid and exudes a foul smell like rancid butter once the seeds have dropped

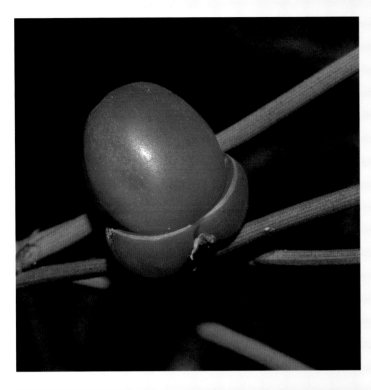

16 spermatozoids from each pollen tube. In more advanced seed plants, such as our conifers, the Gnetales and flowering plants, the two male gametes are reduced to simple cell nuclei, known as *sperm nuclei* or just *sperm*. These sperm can no longer move by themselves, nor do they need to as they are conveniently delivered straight to the egg cell by the pollen tube.

When *mega* really does mean *mega*

Whilst the reduced microgametophytes of the gymnosperms no longer produce the male gametes in specialised containers (antheridia), their female counterparts are true to their name. Not only do the megagametophytes become much bigger, they still go through the troublesome process of forming proper archegonia in which to house their egg cells (except *Gnetum* and *Welwitschia*). Archegonia are always formed beneath the opening of the integument, at the micropylar end of the ovule. How many of them a megagametophyte produces varies greatly within the gymnosperms: the Florida torreya (*Torreya taxifolia*) almost invariably has only one, ginkgos always two, pines usually form two or three, the dawn redwood (*Metasequoia glyptostroboides*) has between eight and eleven, and the Californian redwood (*Sequoia sempervirens*) can bear as many as sixty.

Most cycads produce one to six archegonia per megagametophyte, with one remarkable exception: in the almost extinct *Microcycas calocoma*, called *palma de corcho* (cork palm) in its native Cuba, the female gametophyte produces more than a hundred archegonia.[4] Only five or six of them are actually functional, but they produce the largest known egg cells in the plant kingdom (3mm in length).

Two trees from a single seed?

In the presence of several archegonia, all egg cells are usually fertilised and begin to develop into embryos. In the end, only one embryo survives while the others are aborted. Sometimes, though, more than one embryo matures. Three to four per cent of pine seeds, for example, contain two or even three embryos and when raising pine trees from seed it is possible, in theory, to fine oneself more seedlings than seeds.

The embryo of the gymnosperms

The embryos of the gymnosperms develop in strange ways. The best known example is the pine. In pines (*Pinus* spp., Pinaceae), the fertilised egg cell initially produces sixteen cells near the bottom end of the archegonium. These sixteen cells are arranged in four tiers of four cells each. What happens next is rather bizarre. Each cell on the top layer facing into the megagametophyte gives rise to an embryo, whilst at the same time the four corresponding cells in the second tier underneath produce a stalk-like *embryo carrier* or *suspensor*. The suspensors grow much longer to push the four developing embryos through the wall of the archegonium and into the nutritious tissue of the megagametophyte. Eventually, only the strongest embryo that has been pushed farthest into the megagametophyte survives. The remains of the suspensor can be seen as a thin white thread attached to the root tip of the embryo in a carefully sectioned pine kernel.

Although three of the four embryos produced by one zygote usually die, in rare cases more than one embryo survives. The pine life cycle therefore offers two possibilities for the formation of multiple embryos, a phenomenon called *polyembryony*. If a pine seeds gives birth to two seedlings, it is not possible to know whether the twins are genetically identical.

If conifers are strange, Gnetales are mad

If conifers seem strange, Gnetales are even more mysterious. Their inner integument forms a long, beak-like, tubular micropyle that sticks out beyond the bracts surrounding the ovule. Unusual as their micropyle may be, pollen is captured in a pollination drop at its tip, just as in other gymnosperms. The real differences that make the Gnetales unique are the very private affairs taking place in the ovule. In the case of the comparatively simple *Ephedra*, as in conifers, the megagametophyte develops from the single surviving megaspore and produces two to three archegonia at the base of a deep pollen chamber. Although all the egg cells may be fertilised, only one develops further. The surviving zygote divides three times mitotically into eight daughter-zygotes, of which all (but usually only three to five) can become embryos. In the related *Gnetum*, things are totally different. To begin with, not only one but all four megaspores originating from the meiotic division of the megaspore mother cell survive and participate in the formation of the megagametophyte. There are no archegonia housing the egg cells. In fact, at the time of fertilization, the female gametophyte consists of just one large cell with numerous free nuclei. Each of these free nuclei can act as an egg cell and, since several pollen tubes can reach the megagametophyte, result in numerous zygotes. These zygotes have stalks (suspensors) that may divide and, once more, produce multiple embryos. As usual, only one embryo per seed reaches maturity.

The love tubes of *Welwitschia*

Gymnospermous sex practices become even weirder with *Welwitschia*, the third and most bizarre of the Gnetales. As in *Gnetum*, the female gametophyte develops from all four megaspores and lacks archegonia. At pollination, it consists of several multi-nucleate cells forming a kind of embryonal mass. Some of the cells of this embryonal mass grow tubes up

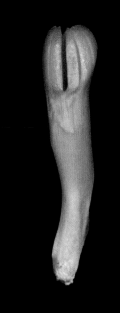

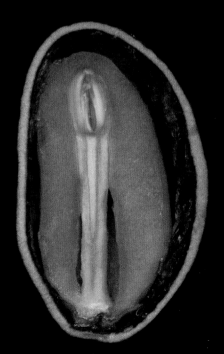

opposite: *Pinus lambertiana* (Pinaceae) – sugar pine; collected in California – longitudinal section of seed (below) and excised embryo (above)

below: *Pinus tecunumanii* (Pinaceae) – piño rojo; native to Central America – seedling showing the four cotyledons surrounding the emerging primary shoot with its numerous young leaves

into the surrounding tissue of the megasporangium where they meet the incoming pollen tubes. The fusion of the sperm with the egg cell nucleus allegedly takes place in these "love tubes" but the developing embryo is pushed back down the tube by its elongating suspensor.

Simple cycads

After so many bizarre ways of forming embryos, the case of the cycads is relatively simple. The zygote produces a single embryo, which is equipped with a long, non-dividing suspensor. In the mature seed, the remains of the suspensor are still attached to the tip of the radicle and appear as a coiled white thread when the germinating embryo exits the seed.

How many leaves on a gymnosperm embryo?

The mature embryo of the gymnosperms typically consists of one or more embryonic leaves (*cotyledons*) attached to the tip of a short straight stem (*hypocotyl*), which has a tiny root (*radicle*) at the bottom. There may be just one cotyledon (in *Ceratozamia*, a cycad), two (most cycads, cypresses, monkey puzzles, yew, *Ginkgo*, *Ephedra*, *Welwitschia* and *Gnetum*), three (*Encephalartos*, another cycad, and occasionally *Ginkgo*, yew and *Gnetum*), four (*Sequoia*), or many as in the pine family Pinaceae (for example, about twelve cotyledons in *Pinus pinea*). In the two weirdest gymnosperms – *Welwitschia* and *Gnetum* – the embryo develops a unique, unparalleled organ – the *feeder* – a lateral outgrowth of the hypocotyl, larger than the embryo itself.

The haploid lunch pack of the gymnosperms

During the development of the gymnosperm seed, the tissue of the female gametophyte nourishes the growing embryo, stocking large amounts of energy-rich fat and protein until it becomes the young sporophyte's lunch pack. If a pine kernel, for example is cut crosswise in half (the hard seed coat has been removed from commercially available pine kernels) in the round pine embryo is in the centre, surrounded by its caring megagametophyte.

The downfall

Gymnosperms and ferns dominated the forests and swamps throughout much of the Mesozoic (including the Triassic, Jurassic and Cretaceous) until about 100 million years ago. Cycads, ginkgos, conifers, and many long lost species were the food plants of dinosaurs. Most are now extinct with less than a thousand species alive today, a shadow of their original diversity. Why? Because of yet another revolution, the advent of "flower power".

Liriodendron tulipifera (Magnoliaceae) – tulip tree; native to eastern North America – the gynoecium in the centre of the flower consists of numerous separate carpels arranged in a spiral around a central axis; in the mature fruit each carpel has developed into a single-seeded winged nutlet (samara)

The Flower Power Revolution

About 140 million years ago, during the late Jurassic (206-142 million years ago) or early Cretaceous (142-65 million years ago) when dinosaurs were in their heyday, a new group of plants appeared that would soon change the face of the Earth: the *anthophytes*. The name may be unfamiliar but anyone who has stopped to admire and smell a beautiful flower knows them. Nearly all plants – grasses, wildflowers, shrubs and trees (except conifers), vegetables, cacti, palms, orchids – belong to them. *Anthophytes* is the Greek for *flowering plants*.

Although named "flowering plants" they were not the first plants to produce flowers. Scientifically, a flower is defined as a specialised short shoot, the growth of which terminates with the production of one or more sporophylls (leaves bearing male micro- or female megasporangia). In terms of growth, a flower is therefore a final act. This is why no shoot ever emerges from the centre of a flower. Strictly speaking, the male and female cones of cycads are male and female flowers. In an attempt to restrict the term *flower* to the structures found in angiosperms, the preferred definition requires the presence of additional sterile (non spore-producing) floral leaves surrounding the sporophylls, for example, the petals of a rose. However, even this does not secure the angiosperms the exclusive right to the use of the word *flower*. Small, not very showy flowers with peripheral leaves around the sporophylls exist in the gymnospermous Gnetales (*Ephedra*, *Gnetum* and *Welwitschia*). The name *anthophytes* – or *flowering plants* – is therefore not the most appropriate for this group of plants. To reflect their revolutionary innovation and to exclude flower-bearing gymnosperms, they are better addressed as *angiosperms*.[5] *Angiosperm* is the counterpart to the term *gymnosperm* (*naked seed*) and means *covered* or *enclosed seed*. But how and why did the angiosperms cover their seeds?

The evolution of the carpel

Take a sporophyll of the primitive gymnospermous sago palm (*Cycas revoluta*, Cycadaceae). It is simply a leaf with ovules lined up along its periphery. Now imagine that this sporophyll folds up along the midrib and its opposite margins fuse to form a bag with the ovules inside. The result is what botanists call a *carpel*. However, the ovules locked away inside the carpel face a small hurdle. How will the pollen, or at least the male sperm, reach them? This problem was solved by the creation of a pollen-capturing zone, the *stigma* (Greek for *spot* or *scar*). The stigma is a wet, receptive tissue on the surface of the carpel, initially along the line where the margins of the sporophyll fused but later reduced to a small platform at the tip of the carpel. It provides pollen grains with ideal conditions for germination. Pushing their way into the stigma, the pollen tubes soon reach a special canal or transmitting tissue that supports their growth and guides them to the ovules down inside the carpel, the womb of

opposite: *Drimys winteri* (Winteraceae) – Winter's bark tree; native to Central and South America – the gynoecium in the centre of the flower consists of a cluster of separate carpels (apocarpous gynoecium)

below: *Drimys winteri* (Winteraceae) – Winter's bark tree – longitudinal section of a carpel showing a series of suspended ovules

bottom: *Drimys winteri* (Winteraceae) – Winter's bark tree – close-up of gynoecium

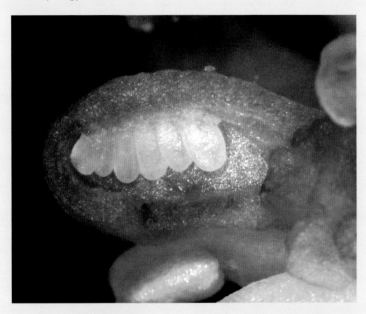

the flower. With the development of the stigma, the initial handicap created by the closed carpel has – quite elegantly – been turned into yet another advantage: whilst the naked ovules of the wind-pollinated gymnosperms have to be pollinated individually, the stigma of the angiosperms has created a single entry point for all incoming pollen. A sole pollination event delivers enough pollen for the fertilization of all the ovules in the carpel. Furthermore, the germination of the "wrong" pollen can easily be inhibited or even prevented by chemical signals produced by the stigma surface.

The carpel was without doubt the revolutionary innovation of the angiosperms and would pave the way for their almost total domination of the plant world. But why is it *such* a great advantage for ovules to be enclosed within carpels rather than "naked" along the leaf margin or on the surface of the cone scales? It is true that they are better protected from predators, but a cone can do more or less the same job and it is much easier for the pollen to get to the ovules if they are not locked away. In order to fully comprehend the significance of the evolution of the carpel, it is necessary to look at the bigger picture.

Gone with the wind

Early seed plants relied on the wind to transport and disperse their pollen in the same way that mosses and ferns entrusted their spores to passing air currents. Wind pollination is still the preferred method among gymnosperms but it is quite an expensive way of getting the pollen to the ovules. There is only a minute chance that the wind will carry a pollen grain straight to an ovule of the same species. Hence, the wind is not a particularly reliable courier. Cycads and conifers make up for this by producing huge amounts of pollen: clouds of yellow pollen can be seen coming from the small male cones of a pine tree in spring when the wind blows. Despite the strategic placement of the larger female cones at the tips of the branches, it takes an enormous number of pollen grains to ensure their successful pollination. When it comes to pollinating flowers, insects are much more reliable and targeted than the wind. Insect pollinators such as bees move from flower to flower seeking rewards, typically in the form of pollen or nectar, and thereby ensure the relatively precise movement of pollen. Insect-pollinated flowers therefore need to produce fewer pollen grains to ensure pollination, a clear reproductive benefit over wind-pollinated flowers. This rather intimate relationship between seed plants and animals took some time to become established.

"Louis, I think this is the beginning of a beautiful friendship"[6]

Animals, birds and insects occasionally visited the flowers of wind-pollinated early gymnosperms. Among these early visitors were insects with strong mandibles (mainly beetles), which were able to chew through the tough sporopollenin wall to gorge on the nutritious contents of the pollen. Thirsty after their meal, they sometimes also visited the female flowers to take a sip from the sugary pollination drop at the tip of the ovule and thus, unintentionally, deposited some pollen. This rather casual relationship gradually developed into something more serious.

Modern-day cycads, for example, are dioecious, which means that they bear their male and female cones (flowers) on separate individuals. When the time for pollination arrives, both male and female cones emit heat and a strong odour that attracts insects (e.g. weevils), as their scales (sporophylls) begin to loosen and separate. This is also the strategy of the cardboard palm[7] (*Zamia furfuracea*), a cycad of the coontie family (Zamiaceae). When mature, the male cones attract swarms of tiny weevils by offering them shelter, food (nutritious pollen) and even a breeding place; but the female cones are poisonous in order to protect the precious ovules. They would therefore have nothing to tempt potential visitors if they did not cleverly trick them by mimicking the appearance and smell of the male cones. This is already quite smart for a supposedly primitive gymnosperm, but it is the much more advanced angiosperms that have become the true masters of animal seduction. The need for less pollen to achieve successful pollination was a great advantage since it meant substantial savings in energy and materials. Moreover, with their ovules safely stowed away in carpels, sufficient safeguard against hungry animal visitors was also in place. Angiosperms therefore very quickly discovered the enormous advantages of a close friendship with insects and other animals, and since this niche was still largely unoccupied, they were able to exploit it relentlessly. How? By means of a beauty contest.

The secrets of attraction

Their newly developed friendship with animals gave rise to stiff competition between the angiosperms for the attention of potential pollinators. In order to become more attractive to catch the eye of passers-by and to make pollination more efficient, angiosperms developed the conspicuous, often colourful structures that are thought of as "true" flowers.

One of the secrets of a proper flower is a successful advertising strategy to lure potential pollinating customers. To make their flowers more conspicuous, angiosperms added colourful leaves to the shoot bearing the sporophylls and often an enticing fragrance. Take the rose. The queen of flowers owes its wondrous beauty entirely to its showy petals, which consist of modified leaves around the reproductive organs in the centre of the flower. Its exquisite scent complements the positive experience, enhancing the attraction of the flower just as an enchanting perfume adds to the allure of a beautiful woman. Another major step forward in

opposite above: *Centaurea montana* (Asteraceae) – mountain bluet; native to Europe

opposite below: *Orbea semota* (Apocynaceae) – native to East Africa – the flower of this succulent is adapted to attract carrion-flies for its pollination; its colouring, the shimmering hairs around the edge, and its smell are all in imitation of a dead animal

below: *Cedrus atlantica* (Pinaceae) – Atlas cedar; native to North Africa – male cones shedding their wind-dispersed pollen

the evolution of the angiosperm flower was the combination of microsporophylls and megasporophylls in a single flower, something only very few gymnosperms (most of which are extinct today) ever managed. Such bisexual or hermaphrodite[8] flowers avoid the duplication of effort required by separate male and female flowers, both of which would have to be equipped with attractants and rewards for pollinators. Since the microsporophylls (stamens) and megasporophylls (carpels) are in the same flower pollen can be received from visiting insects and at the same time some of the flower's own pollen may get attached to the visitors. Bisexual flowers are simply a one-stop shop for receiving and dispatching pollen, as well as for rewarding the dispatcher with food (pollen) and drink (nectar).

Floral architecture

A typical angiosperm flower consists of four or five whorls of specialised leaves. The outer whirl is the *calyx*, a cup-shaped structure formed by three to five small green leaves, called *sepals*. Within the calyx is the large, often brightly coloured *corolla*, usually made up of three to five *petals*. Sepals and petals together form the *perianth* of a flower. Between or opposite the petals, one or two whorls of microsporophylls or *stamens* are inserted. A stamen consists of a slender stalk, the *filament*, carrying the *anther* at the top. The anther is the fertile part of the stamen and bears four microsporangia, the pollen sacs. The stamens themselves encircle the central whorl, the female part of the flower, the carpels. The number of carpels in each flower depends on the species. They can be numerous, as in buttercups (*Ranunculus* spp. Ranunculaceae), the marsh marigold (*Caltha palustris,* Ranunculaceae) and more exotic examples, such as the Winter's bark tree (*Drimys winteri*, Winteraceae) and its relatives, the magnolias (*Magnolia* spp., Magnoliaceae) Other species, such as members of the legume family (Fabaceae), which includes beans and peas, have only one carpel per flower.

Why everything has two names

Scientists have given the sexual organs of seed plants different names although their direct equivalents were already present and properly named in the life cycles of mosses and ferns. In seed plants, microsporophylls and megasporophylls are called stamens and carpels, the microsporangium and megasporangium are the pollen sac and nucellus, and the microspores are pollen grains. It may be true that some scientists love to create new terms from the vocabulary of our Greek and Latin ancestors, but in the case of seed plants at least, it is for strictly historic reasons.

Before Hofmeister's revolutionary discovery in 1851, scientists had already created a whole set of different technical terms for the reproductive organs of seed plants. It was the

Spergularia rupicola (Caryophyllaceae) – rock sea-spurrey; collected in the UK – seed; no obvious adaptations to a specific mode of dispersal but displaying the intricate surface pattern typical of the pink family: the papillose cells of the seed coat are interlocked like pieces of a jigsaw puzzle as revealed by the intricate pattern of undulating lines around the individual cells. Dispersal is achieved when the capsules fling out the seeds as they sway in the wind; the flattened shape of the seeds may assist further dispersal by the wind; 0.6mm long

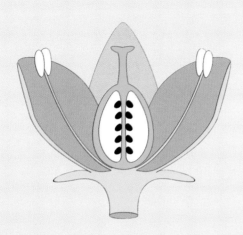

above: Diagram of a typical angiosperm flower (from periphery to centre): floral axis and calyx (bright green); petals (pink); anthers (yellow); ovary with style and stigma (dark green); ovules (black)

below: *Spergularia rupicola* (Caryophyllaceae) – rock sea-spurrey; native to Europe – close-up of flower

English naturalist and physician Nehemiah Grew (1641-1712) who invented much of the terminology used today for the different parts of a flower. Grew began his observations on the anatomy of plants in 1664 and in 1672 he published his first important essay *Anatomy of Vegetables begun*, which was followed in 1673 by *Idea of a Phytological History*. His most important publication on the *Anatomy of Plants* appeared in 1682. It included a chapter on the "Anatomy of Leaves, Flowers, Fruits and Seeds" in which he analysed the function of flowers and for the first time identified stamens and carpels as male and female sex organs. By the time Hofmeister was able to prove that the sex organs of cryptogams (mosses, ferns, etc.) and seed plants have a fundamental evolutionary similarity, Grew's terminology had long been established. Both sets of terms have been retained and are used in parallel.

Fusion technology

During the course of evolution, angiosperms developed a tendency towards fusing parts of their flowers, especially the carpels. In more advanced families, the carpels are usually united to form a single *ovary* or *pistil,* for which the scientific term is a *syncarpous gynoecium*.[9] In a syncarpous gynoecium the carpels share a single stigma, which can be raised above the swollen ovule-bearing part by a slender extension of the carpels, the *style*. The shared stigma helped further rationalise pollination since a single pollination event could now achieve the fertilization of all ovules of not only one but several carpels. This fusion of carpels is easily visible when the ovary of a tulip, for example, is cut in two. Even though they are fused together, the three carpels of a tulip flower retain their walls and divide the ovary into three clearly discernible chambers with ovules inside.

Other parts of the angiosperm flower with a strong evolutionary tendency to amalgamate are the petals. Good examples of this phenomenon are the bellflowers (campanulas) where the five petals form the single, bell-shaped corolla that gave the plants their name. Many evolutionarily advanced families such as the figwort family (Scrophulariaceae) and mint family (Lamiaceae) have flowers with fused petals and mould their corollas into a shape that suits their preferred pollinators, bringing us back to the intimate relationship of angiosperms with animals.

Beauty lies in the eye of the beholder

It was their relationship with animals – especially insects – that permitted the incredible diversity of angiosperms that we marvel at today. The variation of the different flower parts in number, size, shape, symmetry and colour enabled them to develop an almost infinite diversity of flower types. By specialising and tuning their flowers to match the preferences

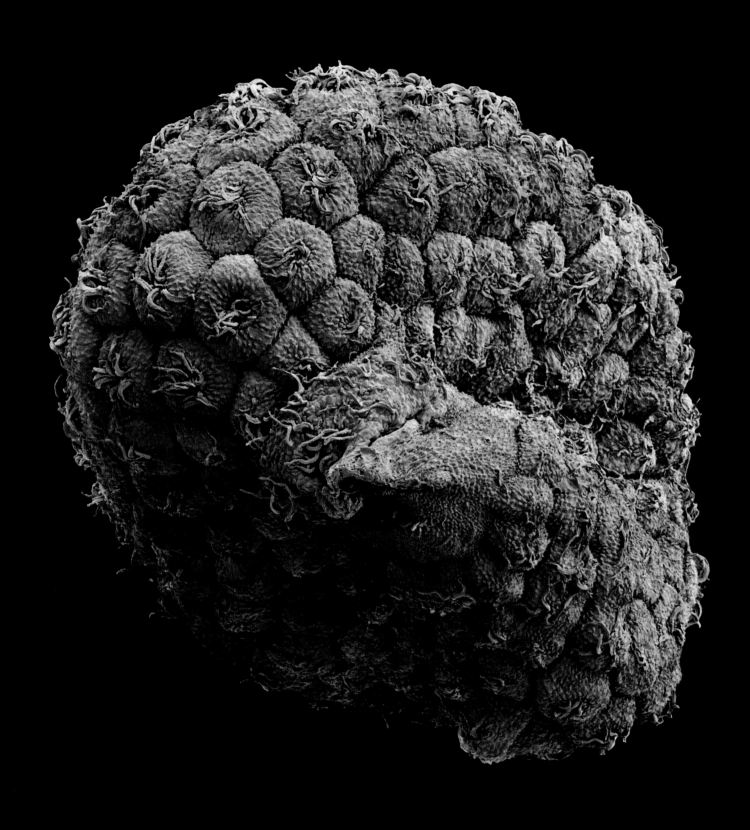

(in colour and smell) and physical needs and skills (body size, length of mouth parts) of only a certain group of insects or sometimes a single species of bee, butterfly, moth or beetle, angiosperms found a very efficient way to avoid unwanted pollen being deposited on their stigmas. This is why some flowers are brightly coloured and emit a pleasant smell (the rose, for example), whereas others are less pleasing to our senses, especially those that are trying to attract carrion flies by looking and smelling like a dead animal (e.g. the smooth carrion flower (*Smilax herbacea*, Smilacaceae) from North America, and African stapelias). Some orchids go as far as interfering with the sex life of their pollinators and mimic a potential mating partner (e.g. the bee orchids of the genus *Ophrys*), or – possibly even more upsetting for the animal – a male rival that must be attacked (e.g. *Oncidium planilabre*).

Quid pro quo *or* the miracle of co-evolution

During the course of evolution, the relationship between plants and insects developed into a very close partnership. In fact, their alliance has become so important for both of them that not only did the plants adapt to the needs of the insect, but the insects also adapted to their flowers. This so-called *co-evolution* between insects and plants was probably the most influential factor in the origin and diversification of the angiosperms.

This mutual adaptation of flowers and insects can be so obvious that Charles Darwin was able to predict the pollinator of the Malagasy comet orchid, *Angraecum sesquipedale*, without having seen it. When he observed the huge 30-35cm long hollow spur inserted in the back of the flower, he postulated that there must be an insect with a tongue long enough to reach the nectar at the end of the spur, most probably a moth. It was only several decades after his death that he was proved right: a giant hawkmoth with a tongue more than 22cm long was captured in Madagascar in the early twentieth century. This long-elusive animal was given the Latin name *Xanthopan morganii praedicta* (*praedicta* meaning *predicted*). Although this hawkmoth had been named and described in 1903, the final proof that it is the pollinator of the comet orchid was not provided until 130 years after Charles Darwin's insightful prediction. In 1992, the German zoologist Lutz Wasserthal went on an expedition to Madagascar to track down the elusive hawkmoth in its natural habitat. The trip was successful: he returned with sensational photographs furnishing irrefutable evidence that *Xanthopan morganii praedicta* is indeed the pollinator of *Angraecum sesquipedale*. This leaves the question of why this hawkmoth developed such preposterously long mouth parts. The answer lies in its feeding strategy. Most hawkmoths feed while hovering in front of flowers. Wasserthal believes that the extreme elongation of the insect's proboscis and the hovering flight are adaptations to protect them from being ambushed by predators such as hunting

opposite: *Scutellaria columnea* (Lamiaceae) – skullcap; native to the Mediterranean – single-seeded nutlet; the ovary of this species is typical of many members of the mint family: it is deeply four-lobed and splits at maturity into four single-seeded nutlets; c.1mm long

below: *Scutellaria columnea* (Lamiaceae) – skullcap; native to the Mediterranean – close-up of flower

stay out of range of hunting spiders. The probable evolutionary scenario is that hawkmoths developed their elongated mouth parts as a defence mechanism. Subsequently, flowers adapted their shape to recruit the suitably pre-adapted hawkmoth as their pollinator.

With its own private courier service in place, a plant species can prevent unwanted hybridization with close relatives. This rather efficient genetic isolation mechanism made it possible for many new species to evolve within a relatively short time, even if they were growing next to their first and second cousins. In addition, a pollinator entrained on a particular flower travels long distances between two flowers of the same species. This allows angiosperm plant communities to be more diverse, which means that they have a higher number of species but a lower number of individuals of each species in a certain space. The best proof that this strategy works is once more provided by orchids. Theirs are the most sophisticated flowers of all angiosperms, and with more than 18,500 species they are the largest and most successful group of flowering plants on our planet. It is their extremely selective pollination mechanism that makes it possible for more than 750 different species of orchids to grow on a single mountain such as Mount Kinabalu in Borneo.

There are always some luddites

As in every society, there are retros among the angiosperms, which never joined the floral beauty contest or dropped out and shifted back to wind pollination. This reluctance or reversal did not occur because the plants found the competition for public attention too stressful. It happened, as so often in evolution, for economic reasons. Wind-pollinated angiosperms are mostly found in places where there are not many pollinating animals but which are windy. In fact, despite the high investment in producing a huge amount of pollen, wind pollination is quite cost-efficient in communities where wind-pollinated plants are common and grow closely together. Good examples are the coniferous forests in the Arctic, the grasslands in Africa but also some angiospermous broadleaf forests in the temperate regions. Deciduous trees like alder, birch, beech, oak, chestnut, hazelnut, and all the grasses are good examples of angiosperms that reverted to wind pollination. Their flowers are small, rather unspectacular (large petals would be an obstacle) and shed huge amounts of pollen into the air in spring – as any hay fever sufferer can testify.

Things can only get better

So far, the angiosperms' most prominent progressive characters (carpels with stigma, bisexual flowers) all revolve *around* the ovules and seeds. With so many major innovations

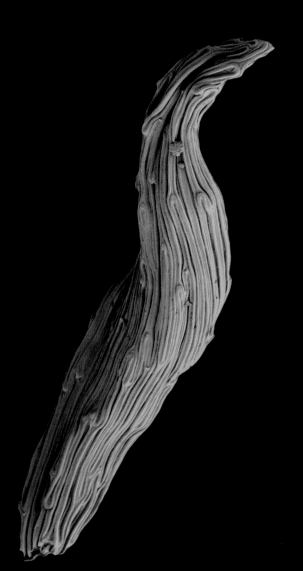

opposite: *Angraecum sesquipedale* (Orchidaceae) – Malagasy comet orchid; native to Madagascar – wind-dispersed balloon seed displaying an elongated, collapsed honeycomb pattern

Pollination of the Malagasy comet orchid (*Angraecum sesquipedale*) by the giant hawkmoth *Xanthopan morganii praedicta*, the only insect with a tongue long enough to reach the nectar at the end of the enormous (30-35 cm long) spur

and a strong tendency towards improving their reproductive organs, it is hard to believe that angiosperms have maintained the same sexual practices as the gymnosperms. But how could they make such an amazingly sophisticated organ as the seed even better? A closer look at the sex life of angiosperms will provide the answer.

The amazing sex life of the angiosperms

If a Kama Sutra had been written for plants, angiosperms would provide the most exquisite example of "vegetable love-making". Their method of sexual reproduction is so sophisticated that, even today, scientists do not fully understand why they do it in the way they do.

Double-wrap protection

Although scientists are not quite sure when, how and why it happened, angiosperms covered their nucellus with a second integument. Such *bitegmic* ovules with an inner and an outer integument are also found among gymnosperms, which are usually *unitegmic* (with only one integument). *Ephedra* and *Welwitschia* are the two gymnosperms with bitegmic ovules; the third member of the Gnetales, *Gnetum* itself, has ovules consisting of a megasporangium surrounded by what some interpret as three integuments. To make matters even more complicated, many angiosperms, especially the most advanced ones, have ovules with only a single integument. However, their unitegmic ovules are derived from bitegmic ones, either by suppressing one integument or by amalgamating the two.

However many integuments an ovule has, there has to be access to the egg cell. In bitegmic ovules, each integument has its own micropylar opening. The one in the outer integument is called the *exostome;* the one in the inner integument the *endostome*. Exostome and endostome together form the micropyle in bitegmic ovules. The area of the ovule opposite the micropyle is the *chalaza*, a topographical term rather than one addressing a specific organ; it refers to the area at the base of the ovule where the integuments join the nucellus.

Umbilical cord and navel

The chalaza is also the place where the vascular supply of the ovule usually terminates. Like a foetus in the uterus, seeds are attached to a *placenta* in the ovary by an umbilical cord. When the seed leaves the fruit, the umbilical cord, called a *funiculus* or *funicle* in plants, detaches from the seed and leaves a scar. This scar, basically the equivalent of the human navel, is present in all seeds, and is called the *hilum*. The hilum is usually small and sometimes almost invisible. In many seeds, however, the hilum is large and constitutes a prominent, characteristic feature of the seed (for example in cacti, horse-chestnuts, certain legumes).

opposite: *Setaria viridis* (Poaceae) – green bristlegrass; probably native to Micronesia but found all over the tropics – fruit (caryopsis) with palea and lemma still attached; c.1.8mm long

below: *Setaria viridis* (Poaceae) – green bristlegrass – fruiting plants

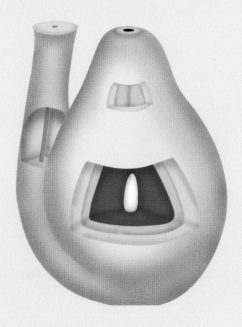

opposite: *Passiflora caerulea* (Passifloraceae) – blue passionflower; native to South America (Brazil to Argentina) – mature ovules suspended from short funiculi, with micropyle pointing towards the fruit wall (placenta); single ovule (without funiculus) 0.5mm long

below: Diagram of a typical angiosperm ovule – light green: outer integument; dark green: inner integument; red: nucellus; yellow: embryo sac; micropyle marked as a black spot at the apex; the tubular structure on the left represents the funiculus (cutaway showing its central vascular bundle)

Upside down

Compared with angiosperms, the ovules of gymnosperms are relatively simple. They have a straight longitudinal axis and the point where they are attached to the mother plant coincides with their chalaza. Enclosed within carpels, the micropyle of such ovules points away from the carpel margin, which is the area from where the pollen tube enters the ovular chamber. In angiosperms, the pollen tube therefore had to grow a certain distance outside the transmitting tissue of the carpel before reaching the micropyle. To solve this problem, angiosperms have turned their ovules upside down so that the micropyle lies closer to the carpel margin. Such "turned" *anatropous* ovules distinguish the angiosperms from the gymnosperms, which have only "unturned" *atropous* ovules.

Whilst hilum and micropyle are located at opposite ends in atropous seeds, they are in close proximity to each other in anatropous seeds. Another indicator of anatropy is the presence of a seam-like structure called the *raphe*. In the mature seed coat, the raphe often appears as a darker or brighter longitudinal ridge or groove extending between opposite ends of the seed. Originally, the raphe was thought to be the portion of the funicle that fused with the integument after its inversion. Although it is true that the raphe is formed by the funicle, it is not really the result of a fusion of tissues: rather it develops from an elongation of the funicle at the point where it is attached to the chalaza of the young ovule. Through this intercalary growth of the funicle the nucellus turns by 180 degrees and the micropyle ends up close to the future hilum.

The anatropous ovule is present in 80 per cent of all families, making it the most common type of ovule in the angiosperms. Only a few exceptions from about twenty families are known to have reverted to atropous ovules. This is mainly the case where the ovary contains only a single erect ovule whose micropyle faces or even touches the transmitting tissue (e.g. in Juglandaceae, Piperaceae, Polygonaceae, Myricaceae, Urticaceae), making it unnecessary to turn the ovule round.

Precocious puberty

Events inside the nucellus of the angiosperms are initially the same as those in the gymnosperms. Within the tissue of the nucellus, one cell, the megaspore mother cell, undergoes meiosis to produce four haploid megaspores of which three (those closest to the micropyle) usually die.[10] The survivor is called the *functional megaspore*.

From this point on, however, everything is different in angiosperms and – most importantly – much faster. Instead of producing a female gametophyte that consists of thousands of cells and one or more archegonia, the megagametophyte of the angiosperms

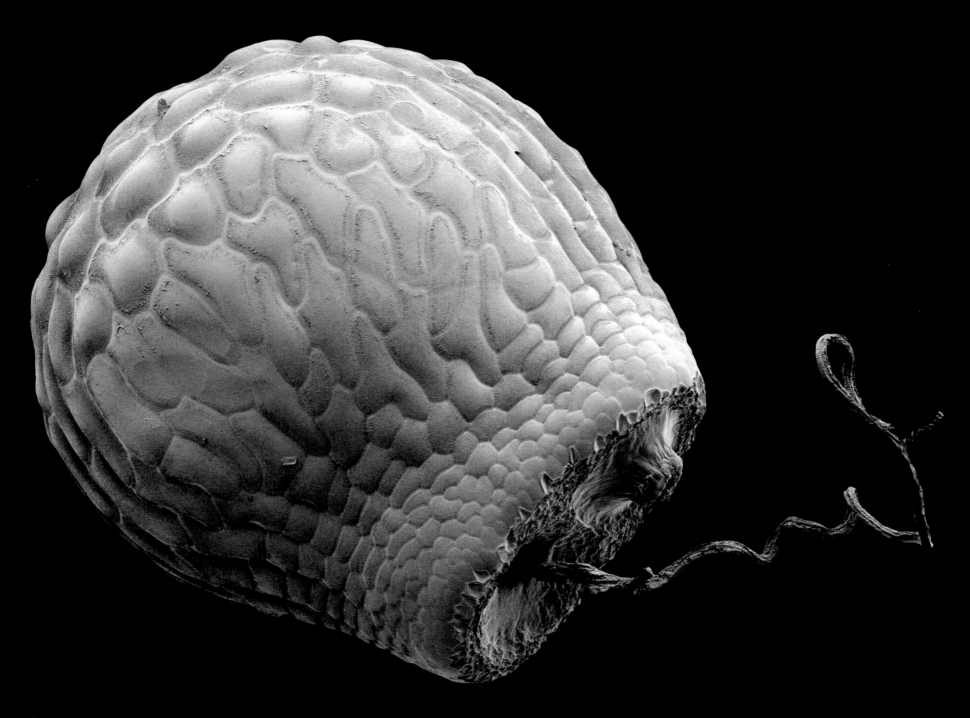

Melocactus zehntneri (Cactaceae) – turk's-cap cactus; native to Brazil – seed with large hilum (brown) covering the micropyle (funnel-shaped protrusion with red rim) and with long central vascular bundle of the funiculus still attached; produced in a berry and dispersed by the animals that eat it, the seed shows no obvious adaptations to a particular mode of dispersal other than a rather smooth surface that ensures easy passage through the gut is simplified. After only three mitotic (normal) divisions of its nucleus, the functional megaspore gives rise to eight free haploid nuclei, which are arranged in two groups of four at either pole of the megaspore. In the next stage, one nucleus from each group migrates to the centre. As they come from opposite poles, they are appropriately called *polar nuclei*. After the formation of cell walls, the mature gametophyte typically consists of eight nuclei distributed over just seven cells, one of which functions as an egg cell.[11] These seven cells of the megagametophyte, in angiosperms also called the *embryo sac*, are arranged in a particular pattern: there are three small cells at the micropylar end forming the *egg apparatus*, which consists of the egg cell and two accompanying *synergids*. The three cells of the egg apparatus face another set of three *antipodal cells* at the opposite end of the embryo sac. Finally, in between the egg apparatus and the antipodal cells is a large *central cell* containing the two polar nuclei. Once the female gametophyte has reached this stage, it is ready for fertilization.

The progressive reduction of the female (and male) gametophytes observed in seed plants is an impressive example of a recurring phenomenon in the evolution of both plants and animals. It is called *progenesis*. Progenesis is the politically correct term for something very similar to "precocious puberty" and refers to an organism that reaches sexual maturity while still in its juvenile stage. Certain amphibians and insects are good examples.

When it comes to sex, angiosperms want it all

The evolution of the seven-celled/eight-nucleate megagametophyte is new and radical, but it is only the foreplay to a most sophisticated act of sexual reproduction. To begin with, gymnosperms have nothing that would compare to the antipodal cells and the polar nuclei of the angiosperms. The antipodal cells usually have no particular function and soon degenerate but the polar nuclei assume a unique role: they join the unification of male and female. But in what way and why are they allowed to join in this most private affair?

On the male side, the microspore starts off as a single cell. Soon, the young microspore undergoes a mitotic division inside the pollen grain, resulting in a vegetative cell and a generative cell. Even before the pollen grains are released from the pollen sacs, the generative cell often divides mitotically to produce the two sperm nuclei. The sperm nuclei are gametes with only a small volume of cytoplasm around the nucleus and no flagellae. When a pollen grain reaches the stigma of a flower, it germinates with a pollen tube. Its growth directed by the nucleus of the vegetative cell, the pollen tube penetrates the stigma and traverses the style, finally entering the cavity of the ovary and eventually the micropyle of the ovule.

Upon entering the micropyle, the pollen tube releases the two sperm nuclei into one of the synergids next to the egg cell. Shortly before this, the two haploid polar nuclei in the

Longitudinal sections of the main types of angiosperm ovules with the embryo sac in red: (a) atropous ovule with two integuments (bitegmic); hilum and micropyle are opposite each other; there is no raphe; (b) anatropous ovule with two integuments; hilum and micropyle are in close proximity; the long raphe extends along the left side of the ovule; (c) anatropous ovule with a single integument (unitegmic)

central cell have fused into one diploid nucleus and migrated to a position close to the egg apparatus. What happens next is unique in the plant kingdom.

Once pollinated, twice fertilised

Whilst their old-fashioned gymnosperm brethren use only one of the two sperm nuclei to fertilise an egg cell and allow the other one to either go to waste or fertilise a neighbouring archegonium, angiosperms want both – to achieve an extraordinary double fertilization: one of the sperm nuclei enters the egg cell as usual, while the other one moves down into the central cell where it meets the diploid nucleus (formed by the two polar nuclei) already waiting close by. Then, in the same way as the first haploid sperm nucleus fuses with the haploid nucleus of the egg cell to form the diploid nucleus of the zygote, the second haploid sperm nucleus fuses with the diploid nucleus of the central cell to form a triploid nucleus. Only when this double fertilization has been successful, will the zygote give rise to the embryo and the ovule start developing into the seed. Before a closer look at the development of the embryo, the fate of the enigmatic triploid central cell should be explored.

The triploid lunch pack of the angiosperms

Even before the zygote produces a recognizable embryo, the triploid central cell grows into a tissue which constitutes the defining component of the angiosperm seed, the *endosperm*.[12] The endosperm surrounds the growing embryo and nourishes it during its development. As the seed matures, the embryo grows deeper into the endosperm. What the embryo has not consumed by the time the seed is mature persists in the seed and serves as the young sporophyte's food reserve during germination. Since it forms the main constituent of cereal grains, endosperm is the reason that rice, wheat, maize, oat and millet provide the staple food for billions of people worldwide. It is this unique and unparalleled triploid seed storage tissue that more than anything else characterises the angiosperms. But why does it emerge from a second fertilization when it does not produce an embryo?

Is the endosperm a sacrificial twin?

In 1898-99 the Russian biologist Sergei Gavrilovich Navashin (1857-1930) and the French botanist Jean-Louis-Léon Guignard (1852-1928) discovered – independently – the developmental origin of endosperm from double fertilization. But since then, the evolution of the specific events that led to the formation of a triploid endosperm has remained a mystery.

Recent research has shown that members of the gymnospermous Gnetales – believed by some to be the closest living relatives of the angiosperms – also undergo a process of

Angiosperm life cycle – clockwise from top: (a) longitudinal section of flower and detail of ovule showing the nucellus (bright green) with the megaspore mother cell (yellow) inside; (b) after meiosis the megaspore mother cell has formed a linear tetrad of megaspores of which only one, the functional megaspore, survives; (c) megaspore after two mitotic divisions containing four nuclei; (d) mature megagametophyte (embryo sac) consisting of eight nuclei distributed over seven cells arranged as follows: three cells at the micropylar form the egg apparatus consisting of the egg cell (orange) and two synergids (white); three antipodal cells (purple) at the opposite end; one large central cell with two polar nuclei (blue); (e) cross-section of anther showing the four microsporangia (pollen sacs) that contain the microspore mother cells; (f)-(j) the microspore mother cell undergoes meiosis to form four pollen grains, each containing one vegetative nucleus (brown) and one generative nucleus (red); (k) pollen tube reaching the ovule in the ovary through its micropyle; in the meantime the generative nucleus has undergone mitosis to form the two sperm nuclei (red); (l) the pollen tube has discharged the two sperm nuclei (red) into one of the synergids (white); one sperm nucleus enters the egg cell (orange), the other joins the two polar nuclei (blue) in the central cell; (m) developing seed with embryo (orange), endosperm (dotted) and the remains of the nucellus (bright green); (n) mature seed; (o) germinating seed; (p) seedling

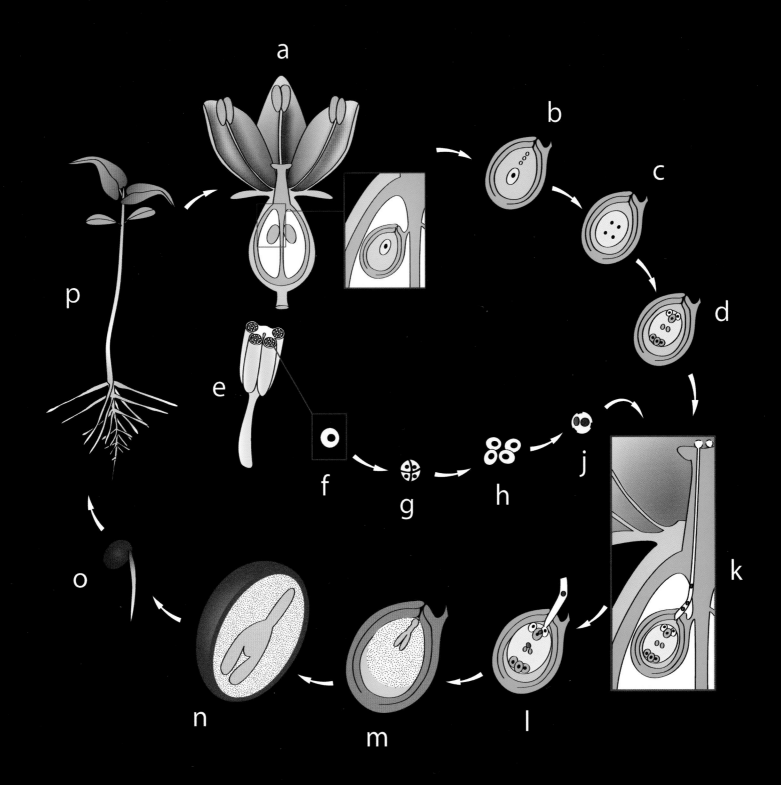

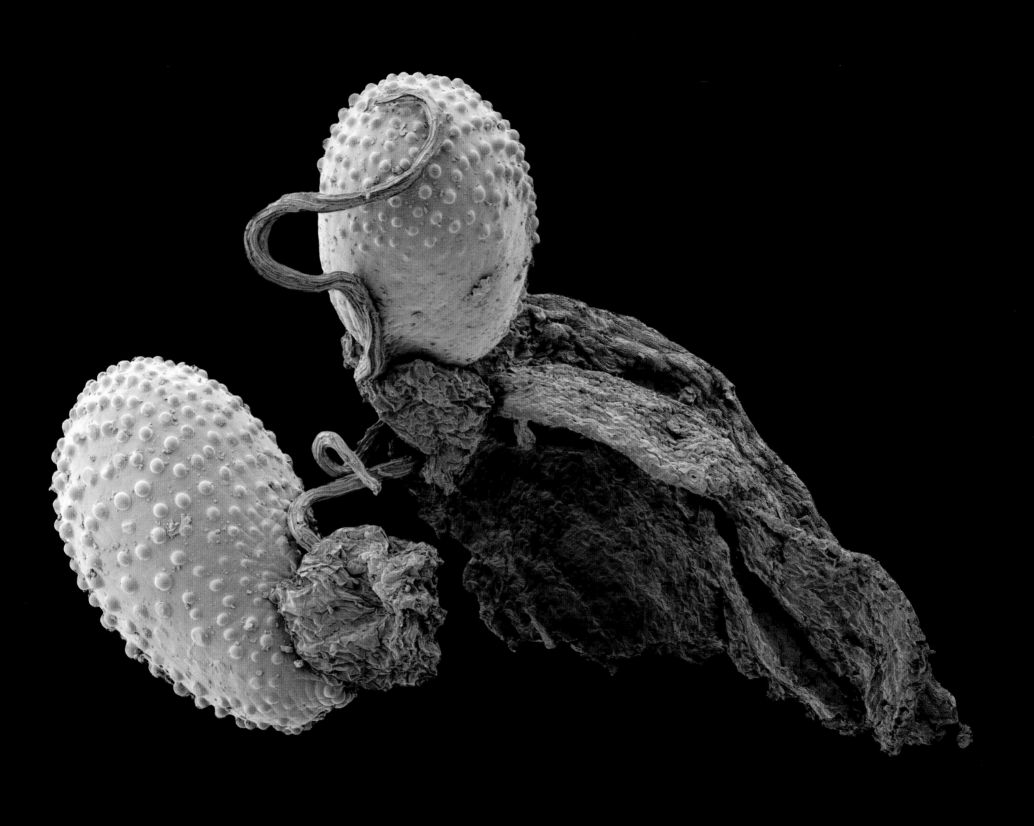

Seeds: Time Capsules of Life

double fertilization. In Gnetales, however, double fertilization does not lead to the formation of a zygote and an endosperm but – as would be expected – to two diploid zygotes. One theory is that in the long extinct ancestors of the angiosperms, the second fertilization event yielded a genetically identical twin embryo but once the angiosperm stem lineage branched off, this twin embryo evolved into an embryo-nourishing structure, the endosperm.

However, it is still not proven that the Gnetales are the closest living relatives of the angiosperms. Recent evidence suggests that they may be more closely related to pines. Their method of sexual reproduction may not, therefore, be of immediate relevance to understanding how double fertilization evolved in angiosperms. Whether scientists will ever be able to solve this riddle is uncertain. Because of incomplete fossil records, angiosperms seem to have appeared suddenly and with considerable diversity in the Earth's history but without obvious antecedents. The evolutionary origin of the angiosperms – the most prominent unresolved issue in plant evolutionary biology – therefore remains Darwin's "abominable mystery" of more than a century ago.

The advantages of double fertilization

Just as the evolutionary origin of the endosperm remains a mystery, the advantages of double fertilization are not completely clear. By incorporating both maternal and paternal genes, the endosperm tissue becomes genetically identical to the embryo. This could enhance the development of the embryo by reducing the risk of genetic incompatibilities between mother and offspring. This is also a possibility in gymnosperms, where the tissue surrounding the embryo (the haploid megagametophyte) is entirely maternal.

From an economic point of view also, the evolution of endosperm seems to be an improvement. Although gymnosperms such as ginkgos, cycads and conifers can be proud of their seeds, the way they "make" them is not particularly efficient. Their primeval sex life forces them to produce their storage tissue (the massive megagametophyte) in advance, before the egg cell is fertilised. Angiosperms put their energy into the formation of the expensive, energy-rich endosperm only after successful fertilization of the ovule. If the pollination of a flower fails, angiosperms have not wasted too much precious energy and materials. Conservation is always a great advantage in the evolutionary race.

The embryo of the angiosperms

After fertilization, or sometimes even before, the synergids and antipodal cells degenerate while the zygote starts to divide mitotically. In all angiosperms, the very first division of the zygote is asymmetrical. It divides the zygote transversely into a large *basal cell* facing the

Glinus lotoides (Aizoaceae) – lotus sweetjuice; collected in Burkina Faso – two seeds still attached to a piece of the fruit wall; the peculiar funicular aril (elaiosome) consists of two lateral lobes with a long tail-like projection between them. The function of the aril is most probably to facilitate dispersal by ants, but it could also assist wind dispersal

mines the polarity of the embryo. The apical cell grows into the embryo proper whereas the larger basal cell produces a stalk-like suspensor, similar to the one already encountered in gymnosperms. Originally, the suspensor was merely conceived as a means to anchor the embryo at the micropyle while it pushed the embryo deeper into the endosperm. Today, we know that its function is much more complex. The suspensor not only nourishes the young sporophyte with nutrients transferred from the mother plant but also controls the early stages of development of the embryo by supplying it with hormones. Unlike the embryo, the suspensor is short lived. In the mature seed it has long since disappeared, often without trace.

Monocots and Dicots

Initially the embryo proper is just a globular lump of cells but it soon starts to differentiate into the embryonic axis (hypocotyl), with the root (radicle) at one end and the first leaves of the young sporophyte, the *seed leaves* or *cotyledons*, at the other. Since the development of the embryo starts right underneath the micropyle, the tip of the radicle always marks the spot.

The embryo of gymnosperms can have any number between one and more than ten cotyledons, but the embryo of angiosperms has either two or just one. Botanists have long used this very convenient, clear-cut distinction to separate the angiosperms into two groups, the Dicotyledons with two seed leaves and the Monocotyledons with only one seed leaf. There are very few exceptions to this rule where the number of cotyledons exceeds two. Some individuals of *Magnolia grandiflora* (Magnoliaceae), for example, may occasionally develop three or more cotyledons, while other species regularly produce three (e.g. *Degeneria vitiensis*, Degeneriaceae) or up to eight cotyledons (*Persoonia* spp., Proteaceae). There are also some Dicotyledons in which the two cotyledons are unequal in size (*anisocotyly*). In extreme cases, such as snowbread (*Cyclamen europaeum*, Primulaceae) and larkspur (*Corydalis* spp., Ranunculaceae), the embryo develops only one cotyledon, while the second one is suppressed. In *Streptocarpus wendlandii* (Gesneriaceae) anisocotyly is not expressed until after germination. In the seed the cotyledons are morphologically identical, but soon after germination one cotyledon dies while the other remains the only leaf the plant will ever develop, eventually becoming as long as 70cm or more. In other Dicots the two cotyledons are fused into one in the mature embryo. Such "pseudomonocotyledonous" Dicotyledons are found in the carrot family (Apiaceae) and the buttercup family (Ranunculaceae). To complete the selection of exceptions, members of the monocotyledonous yam family (Dioscoreaceae) sometimes have two (unequal) cotyledons. The

opposite: *Castilleja flava* (Orobanchaceae) – yellow paintbrush; collected in Idaho, USA – balloon seed displaying the most extreme form of the typical honeycomb pattern found in wind-dispersed balloon seeds: both the outer and inner tangential walls of the seed coat cells dissolve leaving only the loose, voluminous honeycomb cage produced by the thickened radial walls

below: Development of the angiosperm embryo – from left to right): the first division of the zygote is asymmetrical, resulting in a large basal cell and a small apical cell. This first division determines the polarity of the embryo. The apical cell (yellow) develops into the embryo proper whereas the larger basal cell (green) produces a stalk-like suspensor

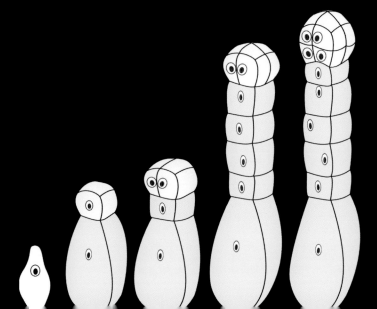

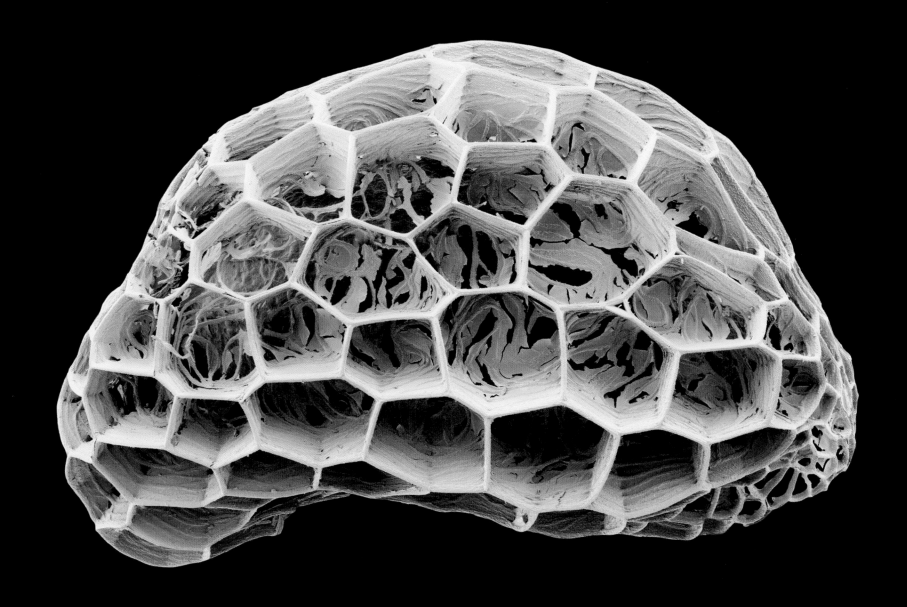

distributed over more than 443 families, compared with 53,000 species in 91 families for Monocotyledons. However, in view of the latest estimate of a total of 422,000 species of angiosperms worldwide these figures are likely to be a gross understatement.

Monocots and Dicots, as botanists abbreviate them affectionately, are generally easy to distinguish, even without counting the leaves of their embryos. Monocots are mostly herbaceous plants with simple leaves that lack a division into a stalk and blade and show parallel venation. All the grasses, bananas, bamboos, lilies, orchids and palms, for example, are Monocots. Recent research has shown that during the evolution of the angiosperms, Monocots branched off as a lineage from within the primitive Dicots. Somewhere on the way they must have lost one of their cotyledons.

Embryo diversity

In many primitive angiosperms, the embryo is very small in relation to the amount of endosperm at the time the seed is released from the mother plant. This is true of members of the annona family (Annonaceae), holly family (Aquifoliaceae), barberry family (Berberidaceae), magnolia family (Magnoliaceae), poppy family (Papaveraceae), buttercup family (Ranunculaceae), and Winter's bark family (Winteraceae). Trying to find the embryo in a seed of a magnolia, custard apple (*Annona cherimola*), poppy, buttercup, Winter's bark tree (*Drimys winteri*), or the twinleaf (*Jeffersonia diphylla*) can be very difficult and seem more like occupational therapy. The embryos in these seeds are microscopically small with hardly discernible cotyledons and the amount of endosperm is huge in relation to the embryo.

A whole range of embryos of different shapes and sizes can be found in seed plants from microscopically small ones to those filling the entire seed. In 1946, Alexander C. Martin published a paper entitled "The comparative internal morphology of seeds" based on an investigation of the seeds of 1,287 plant genera, including both gymnosperms and angiosperms. What Martin found was that the internal structure of seeds varies tremendously with respect to the relative size, shape and position of the embryo. His classification system, which is still widely used today, distinguishes ten different embryo types and two types of extremely small seeds. These twelve types are divided into two divisions. Embryos of the peripheral division are restricted to the lower half of the seed or extend along its periphery. They predominantly belong to the Monocots. Because of the loss of one cotyledon, their asymmetrical embryos can have the most unusual shapes such as flat discs ("broad type") or head-like structures ("capitate type"). Tiny, largely undifferentiated disc-shaped embryos are found in relatives of the grass family (Poaceae)

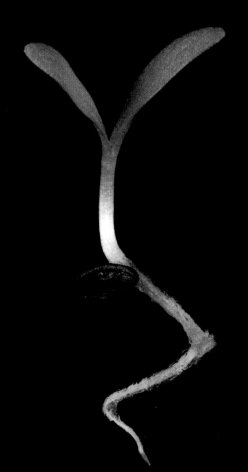

opposite: *Aloe trachyticola* (Asphodelaceae) – seed collected in Madagascar – a typical Monocot seedling displaying the first narrow foliage leaf while the single cotyledon stays inside the seed to resorb the endosperm

below: *Plantago media* (Plantaginaceae) – hoary plantain; collected in the UK – seedling displaying the pair of opposite cotyledons typical of the Dicots

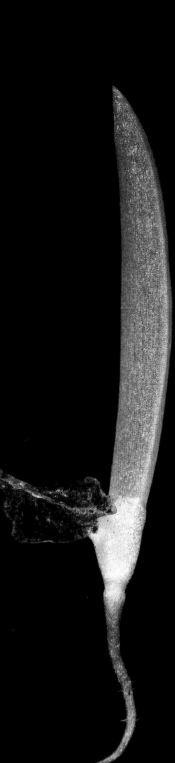

such as the pipewort family (Eriocaulaceae), the rush family (Juncaceae), the restio family (Restionaceae) and the yellow-eyed grass family (Xyridaceae). The rarer capitate embryos are found only in a few families, such as the sedge family (Cyperaceae), the yam family (Dioscoreaceae) and the spiderwort family (Commelinaceae). "Lateral type" embryos, which are in the shape of a wedge tightly pressed against the side of the endosperm, are unique to the grass family. The fourth and last embryo type of the "peripheral division" is (somewhat redundantly) called "peripheral type". It is found in only one order (a taxonomic rank above the family level) of the Dicotyledons, the Caryophyllales, which include the cactus family (Cactaceae), pink family (Caryophyllaceae), pokeweed family (Phytolaccaceae), four o'clock family (Nyctaginaceae), stone plant family (Aizoaceae), purslane family (Portulaccaceae), amaranth family (Amaranthaceae), and knotweed family (Polygonaceae). Here, the embryo occupies a unique position in that it extends along the periphery of the seed rather than following its longitudinal axis through the centre. The reason for the marginal position of the embryo lies in the peculiar nature of the seed storage tissue of the Caryophyllales. Rather than representing triploid endosperm, the central starch-laden tissue around which the embryo curves, originates from the diploid nucellus. Although rare, such a storage nucellus or *perisperm* is found elsewhere in the angiosperms. Most of what is inside an ordinary peppercorn (*Piper nigrum*, Piperaceae) or the seed of a waterlily (*Nymphaea* spp., Nymphaeaceae) consists of perisperm. In the complicated seeds of the ginger family (Zingiberaceae), the storage tissue consists of a significant amount of both endosperm and perisperm.

In seeds belonging to Martin's "axial division" (originally called "axile" by Martin), the embryo follows the longitudinal axis of the seed through its centre. "Linear type" embryos are cylindrical with the cotyledon(s) not significantly wider than the embryo axis. This rather unspecific embryo shape is found in both gymnosperms (conifers, cycads, *Ginkgo*) and angiosperms (both Mono- and Dicots). Restricted to dicotyledonous angiosperms are embryos with broad and flat ("spatulate type") or folded ("folded type") cotyledons. The seeds of the castor oil plant (*Ricinus communis*, Euphorbiaceae) are a good example of the former, and the seeds of cotton (*Gossypium herbaceum*, Malvaceae) of the latter type. Two other types are limited to Dicots: spatulate embryos bent like a jack-knife ("bent type"), and straight spatulate embryos in which the thick cotyledons overlap and encase the short embryo axis ("investing type"). Examples of both types can be found in the legume family (Fabaceae), where the papilionoid subfamily (Papilionoideae) is characterised by the "bent type" and the mimosoid and caesalpinioid subfamilies (Mimosoideae and Caesalpinioideae) typically have "investing type" embryos.

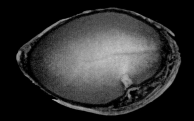 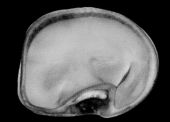 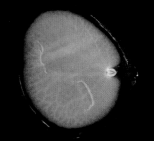

Peripheral, bent and sometimes folded embryos are the result of a curvature of the longitudinal axis of the seed. Seeds with a curved longitudinal axis, where micropyle and chalaza are not opposite each other, are called *campylotropous*. The advantage of campylotropous seeds is that they allow the embryo to become much longer than the actual seed and thus give rise to a taller seedling with improved chances when competing with other seedlings for light.

For very small seeds Martin created two categories based solely on size: seeds that are 0.3 to 2mm long belong to the "dwarf type"; seeds less than 0.2mm long are the "micro type". Martin's measurements exclude the seed coat. Such minute seeds contain tiny, underdeveloped embryos with either poorly developed cotyledons or no cotyledons at all. They are usually wind-dispersed and found in a variety of angiosperm families, most famously in orchids (Orchidaceae) but also in the broomrape family (Orobanchaceae), sundew family (Droseraceae), bellflower family (Campanulaceae) and gentian family (Gentianaceae). Medium or larger seeds with tiny embryos, such as those of the annona family (Annonaceae), magnolia family (Magnoliaceae), buttercup family (Ranunculaceae) and Winter's bark family (Winteraceae), represent Martin's "rudimentary type".

Size does matter

The size of the embryo in the seed is an important factor in the life of a plant. Seeds with very small embryos often need some lead time before they can germinate. During this lead time, which can last for months, the nutrients stored in the endosperm are first mobilised, then absorbed by the embryo. Once the seeds of the ash (*Fraxinus excelsior*, Oleaceae) or those of magnolias have been dispersed, the embryo has to mature and grow larger before it is strong enough to leave the shelter of the seed coat. If seeds with small embryos germinate faster, like those of some palms such as the Mexican fan palm (*Washingtonia robusta*) or the Brazilian needle-palm (*Trithrinax brasiliensis*), the seed remains attached to the seedling until all the endosperm has been resorbed, which takes a long time. The handicap

above: Embryo types (from left to right): *Dypsis scottiana* (Raosy palm, Arecaceae; collected in Madagascar; 12mm long) – rudimentary type; *Anagyris foetida* (stinking bean trefoil, Fabaceae; collected in Saudi Arabia; 16mm long) – bent type; *Brachystegia utilis* (njenje, Fabaceae; collected in Malawi; largest diameter: 5.8mm) – investing type; *Spermacoce senensis* (Rubiaceae; collected in Malawi; 3.8mm long) – linear type; *Porlieria chilensis* (guayacán, Zygophyllaceae; collected in Chile; 9.3mm long) – spatulate type; *Ipomoea pauciflora* (Convolvulaceae; collected in Mexico; 5.1mm long) – folded type

below: Diagram showing the twelve seed types proposed by Martin in 1946 (from left to right and top to bottom; embryo black, endosperm yellow): peripheral division (first row) comprising the basal, capitate, lateral and peripheral types; axial division (second and third row) comprising the linear, dwarf and micro (with and without endosperm), spatulate, bent, folded and investing types; far left in the bottom row is the rudimentary type

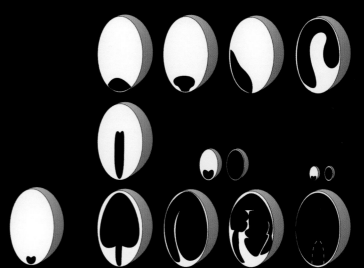

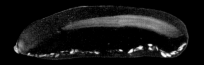 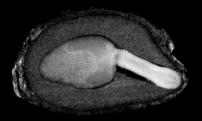 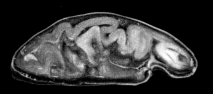

Orobanche sp. (Orobanchaceae) – broomrape; collected in Greece – dwarf type seed displaying the typical honeycomb pattern of wind-dispersed dust seeds; 0.35-0.4mm long

of seeds that are not capable of germinating quickly is that they are unable to react rapidly to small windows of opportunity, such as a sudden downpour in an area of low rainfall (for example, deserts and semi-deserts). The compelling advantages of fast-germinating seeds were therefore a driving force towards the evolution of seeds with larger and more developed embryos. In the most extreme case, the embryo uses up all available endosperm before the seed is mature. The nutrients provided by the mother plant are then stored directly in the embryo's own tissue, usually its leaves. Such storage embryos develop thick and fleshy or thin and folded cotyledons, filling the entire cavity of the seed. With all the nutrients of the endosperm already absorbed before germination, storage embryos are "ready to go" and thus able to take immediate advantage of favourable changes in their environment.

A member of the legume family holds the world record among storage embryos. The endospermless seeds of *Mora megistosperma* (syn. *Mora oleifera*), a large tree from tropical America, can be up to 18cm long and 8cm wide and weigh up to a kilogram, which makes them the largest dicot seeds on earth. The bulk of the seed consists of the two thickened cotyledons as do the seeds of more familiar legumes such as beans, peas and lentils. The only difference is that the seeds of *Mora megistosperma* have an air-filled cavity between the cotyledons, which affords the seeds buoyancy in seawater, an adaptation to their tidal marshland habitat. Other palatable storage embryos with large cotyledons include sunflower kernels, cashew nuts, peanuts and walnuts. That these popular "nuts" are storage embryos explains why they split so easily into two halves, the cotyledons.

The opposite extreme are the dust-like "micro type" seeds of orchids. All they contain inside the wafer-thin loose seed coat is a spherical, underdeveloped embryo without any endosperm. With no significant food reserve for the embryo, orchids have to enter into a symbiotic relationship with compatible mycorrhizal fungi as soon as they germinate. The fungus provides the germinating embryo with precious carbohydrates and minerals. For a few, especially the terrestrial orchids of the temperate zone, the relationship with their

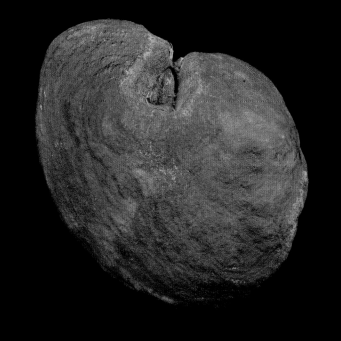
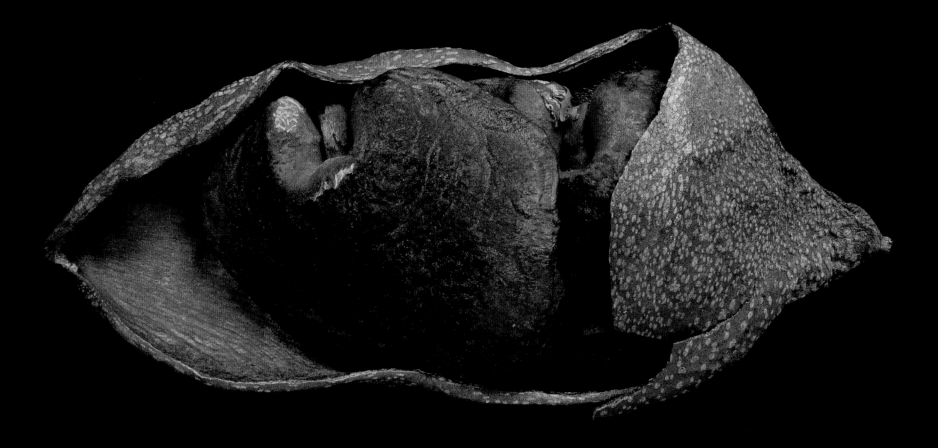

orchids are in choosing their fungal partners. Some orchids at least are known to be able to establish a mycorrhizal relationship with several different species of fungi.

It takes three generations to create one seed

Seeds are composed of the tissues of three different generations; this is true of both gymnosperms and angiosperms. Generation one is the tough protective seed coat. It develops from the integument, which itself is formed by the diploid cells of the parent sporophyte. Generation two is the nutritive tissue, which is provided either by the haploid body cells of the female gametophyte (in gymnosperms) or by the triploid endosperm (in angiosperms). The third generation is the diploid embryo, which combines the genetic material of two different individuals, the mother sporophyte (providing the egg cells) and father sporophyte (providing the pollen). This sophisticated combination of three genetically different generations of tissue in a single organ renders the seed the most complex structure produced by a seed plant.

... or sometimes just one

Although it is true that the vast majority of angiosperms produce their seeds sexually, there are a few notable exceptions. Some plants stopped exploiting the advantages of double fertilization and abstained – either partially or totally – from sexual reproduction. There are more than 400 species in over forty angiosperm families which are able produce seeds asexually, by a process called *apomixis*. Curiously, although well represented among Monocots and Dicots, apomixis appears to be entirely absent among the gymnosperms.

Apomictic species are thought to have evolved independently from sexually reproducing ancestors multiple times. Some of them are only facultative apomicts and can still reproduce sexually, while others are obligate apomicts whose only way of reproduction is apomixis. Although the mechanisms leading to apomictic reproduction are diverse, the underlying principle is that meiotic division is by-passed, so that a diploid egg cell is produced and develops into an embryo without prior fertilization (*parthenogenesis*). For the development of the endosperm most apomicts still require fertilization of the central cell with its two polar nuclei. But some apomictic species have abandoned the fertilization of the central cell, which then initiates the development of the endosperm on its own.

Seventy-five per cent of all known apomicts occur in just three families: the grasses (Poaceae), the rose family (Rosaceae) and the sunflower family (Asteraceae), which include such familiar examples as dandelion (*Taraxacum officinalis*), mouse-ear hawkweed (*Hieracium pilosella*) and cinquefoil (*Potentilla* spp.). As a result of the various apomictic mechanisms the

opposite: *Mora megistosperma* (Fabaceae) – nato mangrove; native to tropical America – individual seed and opened fruit with two seeds; the nato mangrove has the largest seed (up to 18cm long and weighing almost a kilogram when fresh) of all dicotyledonous angiosperms; seed: c.12cm long

below: *Trithrinax brasiliensis* (Arecaceae) – Brazilian needle-palm; native to South America – young seedling with seed still attached; seed 14.5mm long

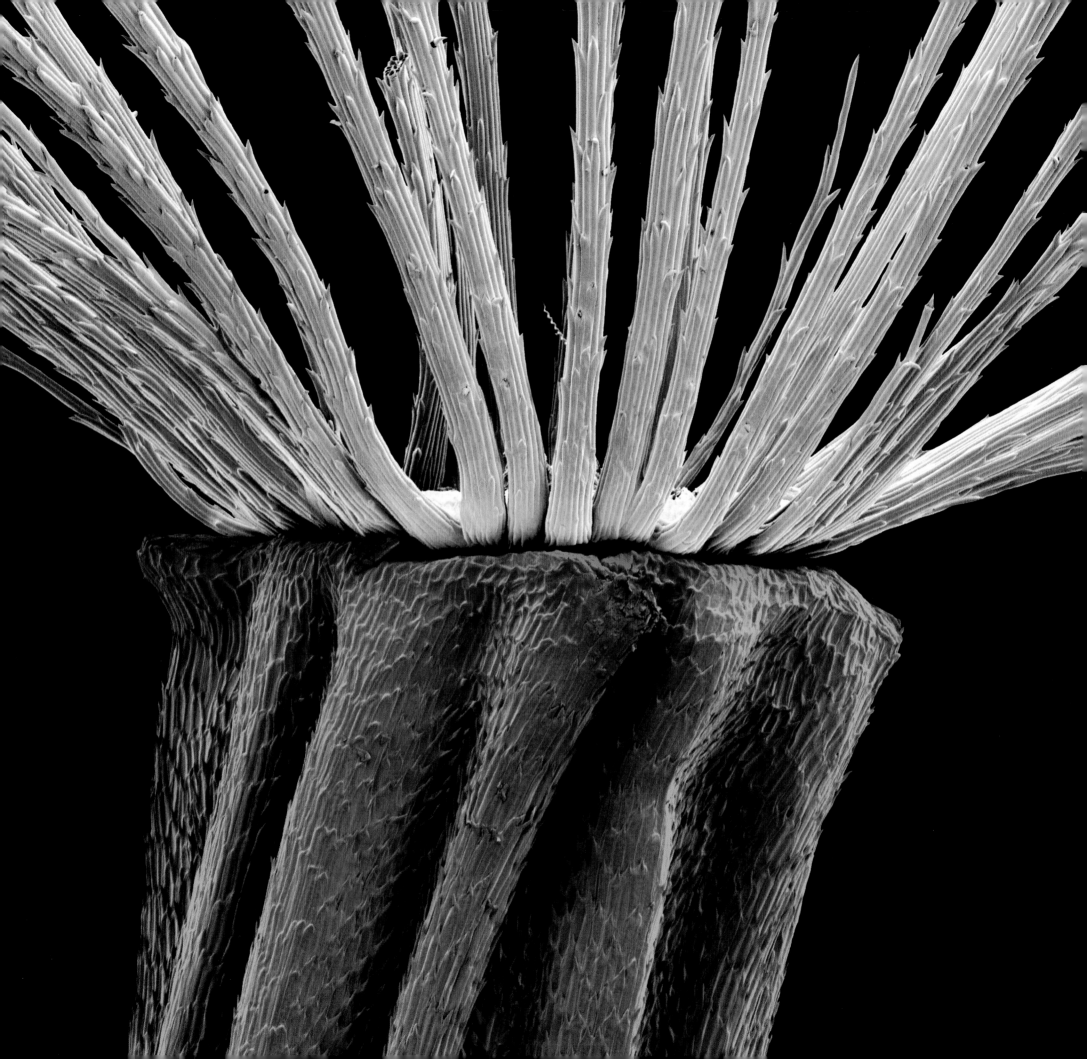

seeds contain embryos that are genetically identical to the mother plant. Such clonal embryos are useful in the propagation of apomictic crops such as *Citrus* species (Rutaceae), mangosteen (*Garcinia mangostana*, Clusiaceae) and blackberries (*Rubus fruticosus*, Rosaceae). Their apomictically produced seeds yield an exact copy of the mother plant, thus perpetuating the valuable traits of a particular race or hybrid through successive seed generations.

Homage to the gymnosperms – at least they look good on paper

Beautiful flowers, clever pollination strategies, double fertilization, and more efficient seed production – the angiosperm way of life was, and still is, a great success story. The fossil record shows that angiosperms appeared quickly and suddenly between the end of the Jurassic and the beginning of the Cretaceous (c.140 million years ago). By the end of the Cretaceous (c. 65 million years ago), they had literally exploded into a huge diversity of species, taking over most terrestrial plant communities. Soon, they had largely displaced ferns and gymnosperms, the dominant vegetation during the Permian, Triassic and Jurassic.

Nevertheless, ferns and gymnosperms still grow on Earth today. In fact, recent research suggests that the majority of living fern species (80 per cent) diversified only in the Cretaceous, after the angiosperms appeared, which is much later than one would expect of such a supposedly ancient group. The new and more complex habitats created by the angiosperms, especially the forests they formed, offered new niches for other organisms. Ferns were probably among the opportunists taking advantage of this, which would explain their Cretaceous comeback. With more than 10,000 extant species they outnumber the gymnosperms ten times. Despite their diminished diversity, gymnosperms still play a significant role today. Conifers in particular put up some serious competition for angiosperms. Although angiosperms mature more quickly than gymnosperms and generally produce more seeds in the same time, conifers, at least, are better adapted to dry, cool habitats. This is why they dominate forests in northern latitudes, at high elevations and on sandy soils. In fact, coniferous forests cover about 25 per cent of the land surface and provide most of the cellulose used in papermaking.

However, with an estimated 422,000 species, it is true that angiosperms vastly outnumber the thousand species of gymnosperms that still exist today. Their incredible versatility and ability to adapt to almost all climates and situations is unrivalled in the plant kingdom and has allowed them to become the unchallenged rulers of the Earth's flora. The fantastic diversity of the angiosperms is displayed in nearly every aspect of their appearance but nowhere more than in their reproductive organs, flowers, seeds and, of course, the containers in which they produce their seeds.

opposite: *Hieracium pilosella* (Asteraceae) – mouse-ear hawkweed; collected in the UK – apex of fruit (cypsela) bearing a large pappus to assist wind dispersal; the seed-bearing part 0.8mm in diameter

below: *Hieracium pilosella* (Asteraceae) – mouse-ear hawkweed; native to Eurasia and North Africa – the mature "seed head", carried on a long upright stalk, presents the tiny parachuted fruits (cypselas) to the wind

Punica granatum (Lythraceae) – pomegranate – longitudinal section of mature fruit, which has a tough leathery skin. White, spongy, membranous walls subdivide the interior into compartments, which contain numerous sarcotestal seeds with transparent, bright red juicy seed coats. The seeds represent slightly more than half the weight of the fruit

below: *Punica granatum* (Lythraceae) – pomegranate; native to Asia (Iran to the Himalayas) and widely cultivated throughout the Mediterranean since ancient times – pollinated flower (the large crinkly red petals have already dropped) displaying the numerous stamens surrounded by the thick, persistent calyx.

bottom: *Punica granatum* (Lythraceae) – pomegranate – young fruit with the large persistent calyx

The seed-bearing organs of the angiosperms

The evolution of animal-pollinated flowers with carpels and more efficient seeds is only the first part of the angiosperm success story. The second part covers the magnificent spectrum of strategies and adaptations developed by angiosperms to disperse their seeds in order to ensure the survival of their species and to expand into new territories. Enclosing the ovules within carpels has many advantages. As the seeds ripen they eventually have to leave the mother plant, and for early angiosperms confinement in the carpel could have been a significant obstacle. Soon though, the initial difficulty of liberating the seeds from the carpels was overcome. One way to solve the problem of angiospermy was to develop one-seededness so that the carpel was incorporated into the seed and both were dispersed together. This condition is still found in most indehiscent fruits such as nuts and drupes. However, during the course of evolution angiosperms have perfected their seed-bearing organs, better known as "fruits", and so turned this initial obstacle into another advantage. They developed a wide range of specific adaptations, which enabled them to exploit every possible means of transport for their seeds and fruits, including wind, water and, most importantly, animals. Once more, the astonishing ability of the angiosperms to diversify and adapt to every available niche contributed towards their evolutionary success. Each dispersal strategy is reflected in a complex syndrome of adaptations that appear after the ovules have been successfully fertilised. It all starts with the wilting of a pollinated flower.

Metamorphosis

As soon as a flower has been pollinated and its ovules fertilised there is no need to attract further pollinators. The showy petals and the stamens usually wilt and wither away, while the ovary starts to swell and turn into a fruit. Inside the ovary, the ovules grow, the endosperm forms, and the embryo develops, while the soft integuments turn into the hard seed coat.

Whereas gymnosperm seeds take at least twelve to twenty-four months to ripen, most angiosperms produce their seeds much faster, usually within a few weeks or months. Bananas, for example, take only two to three months, cherries are edible after three to four months, and mangos can be harvested four to five months (100-130 days) after flowering. Coconuts take a whole year after pollination to mature and Brazil nuts need even longer (fifteen months). With seven to ten years from flower to fruit, the Seychelles nut palm (*Lodoicea maldivica*, Arecaceae) – in many ways an angiosperm extremist – is the slowest fruiter of all. In comparison, the fastest reproducing weeds such as thale cress (*Arabidopsis thaliana*, Brassicaceae) cycle from seed to seed in less than six weeks.

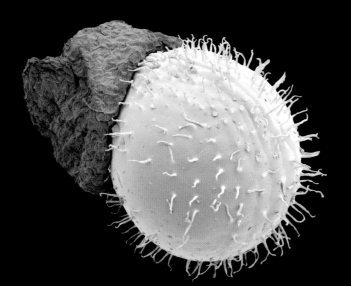

opposite: *Aztekium ritteri* (Cactaceae) – aztec cactus; native to Mexico – seed with funicular aril (elaiosome) that probably assists dispersal by ants. Its extremely slow growth, unusual surface sculpturing resembling Aztec carvings, small size (rarely exceeding 5cm in diameter) and extraordinary habitat (vertical or near-vertical limestone and gypsum cliffs virtually devoid of any other vegetation) make the Mexican aztec cactus one of the most charismatic cacti; seed (including the elaiosome) 0.8mm long

below: *Blossfeldia liliputana* (Cactaceae) – native to Argentina and Bolivia – seed with funicular aril (elaiosome) that probably assists dispersal by ants. The smallest of all cacti with a diameter of only 12mm. Together with the aztec cactus (*Aztekium ritteri*) and the top cactus (*Strombocactus disciformis*), *Blossfeldia* belongs to a group of dwarf cacti naturally restricted to cliffs; they all produce tiny, dust-like seeds 0.65mm long (including elaiosome)

Going to extremes – fruits smaller than a sperm, heavier than a limestone

Although we are sometimes unaware of it, every angiosperm produces some kind of fruit: herb, shrub and tree all produce their own characteristic fruits, and the variety is endless. The spectrum of sizes alone ranges from the almost microscopic fruits of the smallest angiosperm, the aquatic watermeal (members of the genus *Wolffia*), to the enormous fruits of *Cucurbita maxima*, better known as giant pumpkins. To put this into perspective: a full-grown *Wolffia* fruit is just 0.3mm long and smaller than a sperm of the cycad *Zamia roezlii*, whereas the world's largest pumpkins can weigh an incredible 600kg. The angiosperms developed a tremendous diversity of shapes and models, from juicy berries and drupes, hard nuts, winged gliders and helicopters, to exploding capsules and ferociously spiny pods. There are so many different types that botanists over the past two centuries, in a desperate attempt to classify them, generated more than 150 technical fruit names. Explaining the tantalising intricacies of fruit classification is beyond the scope of this book. However, to give an impression of the agony botanists still face, the main criteria that are used to bring some degree of order into the overwhelming diversity of fruit types will be discussed briefly.

A brief introduction to the classification of fruits

Despite the difficulties botanists experience with increasing attention to structural details, the three basic principles of fruit classification are relatively simple: (1) the underlying ovary type (apocarpous or syncarpous); (2) the consistency of the fruit wall (soft or hard); and (3) whether or not a fruit opens up at maturity to release its seeds. Since every combination of characters is possible in angiosperms, all three criteria are applied independently.

The three basic types of fruits

The most important criterion of fruit classification is the structure of the ovary from which a fruit develops. *Simple fruits* like cherries, tomatoes, oranges and cucumbers develop from *one flower* with only *one pistil*, which can be either a single carpel or several united carpels. *Multiple fruits* develop from flowers with *several separate pistils*. Such flowers are typically found in primitive angiosperm families such as the Annonaceae, Magnoliaceae, and Winteraceae. The fruit of the tulip tree (*Liriodendron tulipifera*), a member of the Magnoliaceae family, forms a cone-like structure in which each of the numerous carpels develops into a flat, narrow-winged *fruitlet*. In the Winter's bark tree, *Drimys winteri* (Winteraceae), the apocarpous gynoecium of each flower produces not just one but a whole cluster of bean-shaped berries. A delicious exotic relative of the Winter's bark tree is the custard apple (*Annona cherimola*, Annonaceae). In this case the carpels fuse together as they develop into what appears to be a simple fruit. The only giveaway of the originally apocarpous gynoecium is the distinct hexagonal pattern on the skin of the fruit, marking the boundaries between the individual carpels. More familiar examples of multiple fruits are blackberries (*Rubus fruticosus*) and raspberries (*Rubus idaeus*), both members of the rose family (Rosaceae). Their descent from flowers with apocarpous ovaries explains why they resemble a dense cluster of tiny grapes, each small globule representing a carpel.

The similar-looking mulberries (*Morus nigra*, Moraceae) belong to yet another major category of fruit. The mulberry does not develop from a single flower with several separate pistils but is the joint product of an entire inflorescence (a group of individual flowers). Hence, each bead or *fruitlet* of a mulberry was once a tiny flower. Other culinary examples of *compound fruits* are pineapples (*Ananas comosus*, Bromeliaceae) and the fruits of members of the Moraceae family, such as figs (*Ficus carica*), breadfruits (*Artocarpus altilis*) and jackfruits (*Artocarpus heterophyllus*). Incidentally, a good-size jackfruit can be up to 90cm long and weigh 50kg, which makes it the largest tree-borne fruit in the world.

Drupes and Berries

The two remaining classification criteria separate fruits into fleshy and dry ones, and into those that remain closed (*indehiscent* fruits) or open to release their seeds (*dehiscent* fruits). Although the two criteria are applied independently, soft and fleshy fruits are usually indehiscent and eaten by frugivores, which disperse the seeds in their faeces. The two principal types of fleshy indehiscent fruits are *drupes* and *berries*.

In drupes like cherries, plums, peaches and mangoes, the *pericarp* (the ovary wall in a ripe fruit) is differentiated into three layers: the thin outer *epicarp* (the skin of the fruit); the fleshy *mesocarp* (the actual pulp of the fruit); and a hard inner layer called *endocarp*. Epi- and mesocarp are the edible parts, whereas the endocarp forms the usually single-seeded stone. In berries the pericarp is entirely soft and juicy and remains by definition indehiscent. In contrast with drupes, berries mostly contain several or many seeds. Typical multi-seeded berries are tomatoes, cucumbers, grapes and blueberries. The avocado is an example of a single-seeded berry. As blackberries, raspberries and mulberries have already proven, many fruits we call berries are botanically speaking not berries at all. Juniper berries are the most deceptive example. Famous for giving Dutch genever and gin their characteristic flavour, juniper berries are not berries but the cones of a gymnosperm called *Juniperus communis* (Cupressaceae). In the two to three years during which the female cones of this dioecious conifer ripen, their three uppermost scales develop into a blue, fleshy layer that encloses the seeds almost like the pericarp of a true berry.

Combretum zeyheri (Combretaceae) – large-fruited bushwillow; native to southern Africa – the four-winged wind-dispersed fruits can be up to 8cm in diameter. Hornbills like to eat the seeds from the fallen fruits

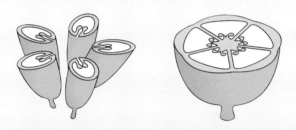

below: cross-sections of an apocarpous gynoecium with five separate carpels (left) and a syncarpous gynoecium with five fused carpels (right)

bottom: *Macadamia integrifolia* (Proteaceae) – smooth-shelled macadamia nut; native to Queensland – fruits; what is generally called a "nut" is the single seed carried inside the leathery green husk (pericarp) of the berry-like fruit; later the pericarp splits along one side but the gap is too small for the seed to pass through

True nuts

Indehiscent fruits with an entirely hard, dry pericarp are called *nuts*. The fruit wall of a nut remains closed, making the independent distribution of several seeds impossible. For the same reason as drupes, nuts are therefore mostly single-seeded. As the above definition indicates, scientists use the term *nut* in a much more precise and slightly different way from everyday language. Botanically, nuts do not include only familiar nuts such as hazelnuts, peanuts and cashew nuts but also "winged nuts" (*samaras*), like those of elms (*Ulmus* spp.) and ash (*Fraxinus* spp.), and much smaller fruits which most people wrongly think of as seeds.

Nuts that are seeds and seeds that are nuts

According to the botanical definition of a nut, the small grains of the grass family (which includes all our cereals) and the "seeds" of the sunflower are, in fact, miniature nuts. Every grain of rice, wheat or oat – small though it may be – is a fruit. Likewise, every sunflower "seed" is a small nut that develops from one of the hundreds of tiny flowers arranged spirally across the disc of the inflorescence. Note that, as is typical for a member of the sunflower family, a sunflower blossom is not a single flower but an entire inflorescence mimicking the looks of an individual flower. Some relatives of the sunflower, like dandelion (*Taraxacum officinale*), meadow salsify (*Tragopogon pratensis*) and mouse-ear hawkweed (*Hieracium pilosella*) attach a kind of parachute (a *pappus*) to their tiny nuts, which allows them to catch a ride on the wind. With all these differences between the various kinds of nuts scientists began to create names for each type. Besides calling winged nuts *samaras*, botanists gave the by now traditional term *caryopsis* to the fruit of grasses, mainly because their thin pericarp is tightly attached to the seed coat. Small nuts with a rather soft pericarp that is separate from the seed coat, such as those of the sunflower, were named *achenes*, or *cypselas* if, like the dandelion, they have a parachute attached.

All types of nuts or nutlets have one feature in common: functionally they act in the same way as seeds. The protective role of the seed coat has been taken over by the hardened fruit wall (pericarp). This explains why, in common parlance, we refer to nuts as seeds and to seeds as nuts. Examples of the latter are treats like (unshelled) pine nuts (*Pinus pinea*, Pinaceae), Brazil nuts (*Bertholletia excelsa*, Lecythidaceae) and macadamia nuts (*Macadamia integrifolia* and *M. tetraphylla*, Proteaceae). While still on the tree, each macadamia seed is wrapped individually in a leathery green husk (the pericarp), which later splits along one side; but the split is much too narrow to allow the seed to escape. Unshelled almonds (*Prunus dulcis*, Rosaceae) and pistachios (*Pistacia vera*, Anacardiaceae) are, in fact, the stones of drupaceous fruits.

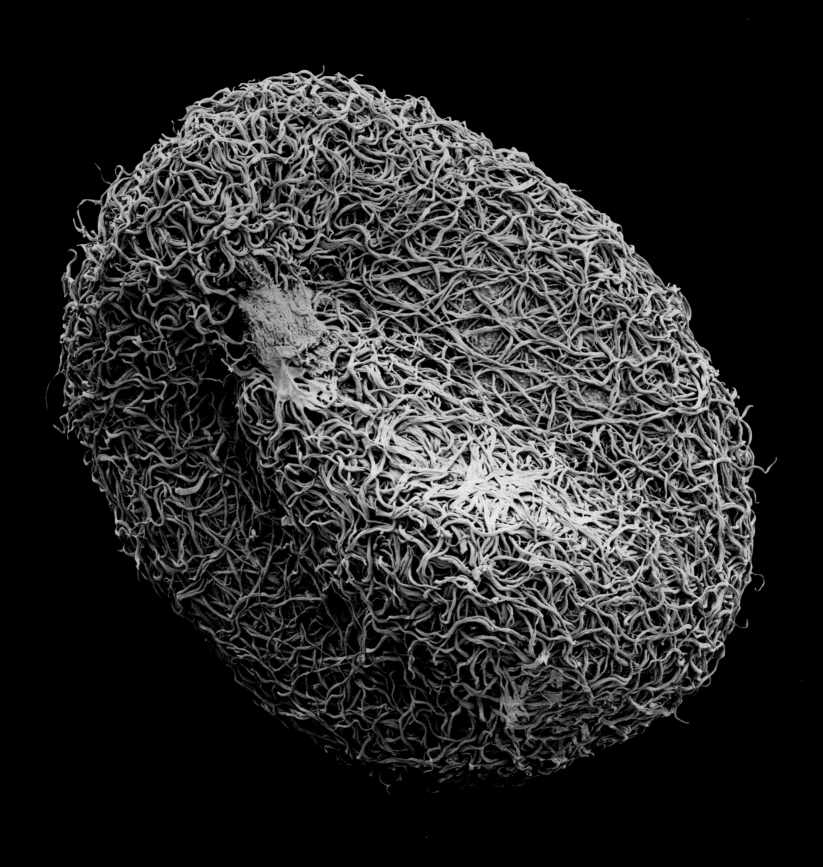

Nutlets

An interesting variation on the simple fruit theme is fruits, which – despite developing from a syncarpous ovary – disintegrate into their carpellary constituents when they are ripe. Such *schizocarpic* fruits break up into two or more *fruitlets*, each fruitlet consisting of either a whole or half carpel. The fruitlets themselves are usually dry, remain closed, and contain only a single seed, which makes them almost real nuts. As they are *fruitlets* rather than *fruits*, they are called *nutlets* instead of *nuts*. Schizocarpic fruits are typical of the carrot family (Apiaceae). Examples of this fruit type can be found on any spice shelf – caraway, cumin, coriander, aniseed and fennel. The winged fruitlets of maple (*Acer* spp., Sapindaceae), which form a pair in the intact fruit, are another familiar example of schizocarpic fruits.

The fruitlets of both the Apiaceae and *Acer* represent a single entire carpel. In other families the division of the fruit goes one step further. The ovary of the mint family (Lamiaceae) and the borage family (Boraginaceae) generally consists of two fused carpels, which are deeply lobed to form twice as many compartments. At maturity, the four compartments separate into single-seeded nutlets, each consisting of half a carpel. Even to the experienced eye, the seed-bearing half-carpels of *Salvia* (sage), *Origanum* (oregano), *Thymus* (thyme) and other Lamiaceae look deceptively like true seeds, which is what most people consider them to be. A bizarre example of a schizocarpic fruit is that of *Ochna natalitia* (Ochnaceae). In the flowers of the genus *Ochna*, the carpels are joined only at the base and share a common style. Once pollinated and fertilised, the carpels develop into separate drupelets that sit, like eggs around the edge of a plate, on an enlarged, fleshy red flower axis.

Capsules

Fruits in which the pericarp opens (dehisces) to expose or release the seeds are called *capsules*. There are several ways in which the pericarp can open: through pores, by a circular split defining a lid, or by regular longitudinal slits.

Poppy capsules, for example, dehisce via a ring of pores around the top. The seeds are ejected through these pores like salt from a saltshaker as the fruits sway in the wind on their long, slender stalks. Examples of capsules that open with a lid are those of the scarlet pimpernel (*Anagallis arvensis*, Primulaceae), the twinleaf (*Jeffersonia diphylla*, Berberidaceae), the squirting cucumber (*Ecballium elaterium*, Cucurbitaceae) and the monkey pot (*Lecythis pisonis*, Lecythidaceae) from tropical South America. However, most capsules split open along regular lines, which either follow the *septae* (the walls between the individual carpels of a syncarpous gynoecium) or run down the middle of each *locule* (the seed-bearing cavity of a carpel). Capsules of the first type are called *septicidal* capsules, those of the second type

Scutellaria orientalis (Lamiaceae) – oriental skullcap; native to southern Europe – single-seeded nutlet; as in other members of this family, the ovary is deeply four-lobed and when mature splits into four single-seeded nutlets, each 1.8mm long

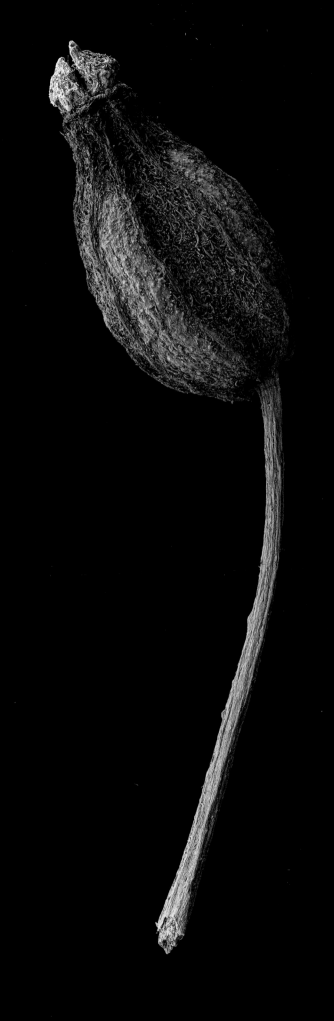

Common spices of the carrot family (Apiaceae) – from left to right: *Pimpinella anisum* (anise; native from Greece to Egypt; fruit 3.5mm long); *Carum carvi* (caraway; native to the Mediterranean; fruit 6mm long); *Cuminum cyminum* (cumin; native to the Mediterranean; fruit 5mm long); *Anethum graveolens* (dill; native to Central Asia; fruit 4.5mm long); the fruits of the carrot family develop from a syncarpous ovary composed of two carpels that separate at maturity to form individual nutlets. In the spices shown the two nutlets are still joined

loculicidal capsules. Septicidal capsules are less common than loculicidal capsules; the foxglove (*Digitalis* spp., Plantaginaceae) and tillandsias (*Tillandsia* spp., Bromeliaceae) are two examples with septicidal capsules. The more common loculicidal capsules are found in many Dicots such as the horse chestnut (*Aesculus hippocastanum*, Sapindaceae), the Himalayan balsam (*Impatiens glandulifera*, Balsaminaceae), and in many Monocots such as irises (*Iris* spp., Iridaceae), alliums (*Allium* spp., Alliaceae) and fritillaries (*Fritillaria* spp., Liliaceae).

Capsules are mostly described as dry fruits but notable exceptions include the fleshy capsules of the Himalayan balsam (*Impatiens glandulifera*, Balsaminaceae), the squirting cucumber (*Ecballium elaterium*, Cucurbitaceae) and tropical hedychiums (*Hedychium* spp., Zingiberaceae). To fully conform to the strict botanical definition of a capsule, a fruit not only has to be dehiscent but must also develop from a syncarpous ovary composed of two or more united carpels. Thus, the fruit of the twinleaf, which as a member of the barberry family (Berberidaceae) has only one carpel per flower, is strictly a *capsular fruit* and not a true capsule. The same applies to the entire legume family (Fabaceae) in which the typical fruit is a single carpel that opens along the dorsal and ventral suture, splitting the fruit in half. Their capsular fruit, familiar from many ornamental (e.g. lupins, sweet peas) and edible plants (e.g. beans, peas, lentils) is traditionally called a *legume*. If a single carpel opens only along one line (usually along the ventral side), as in the multiple fruits of the buttercup family (Ranunculaceae) and star anise (*Illicium verum*, Schisandraceae) or in the cone-like compound fruits of banksias (*Banksia* spp., Proteaceae), it is termed a *follicle*.

Indehiscent capsules

Legumes and follicles earned their own names by not conforming to the definition of a true capsule; they failed on one criterion, the syncarpous ovary. In other cases, many botanists seem willing to turn a blind eye: for example, they refer to the "fruit" of the baobab (*Adansonia digitata*, Malvaceae) as an indehiscent capsule. The large fruit of the baobab tree does give the overall impression of a capsule but it does not open, hence the paradoxical name *indehiscent* (non-opening) capsule. The dry, hard shell of the baobab fruit loosely encloses thirty seeds, which has led to the name "Judas fruit". The seeds themselves are embedded in a white mealy pulp, called "cream of tartar". This pulp is high in vitamin C and used to prepare porridge for newborn babies if the mother has insufficient milk, or it is made into a refreshing drink to treat fever and diarrhoea; it can also serve as a cholesterol-free Parmesan substitute. Another "square peg" in the "round hole" definition of a capsule is the famous Brazil nut tree (*Bertholletia excelsa*). Its large woody fruits take fourteen months to mature and their pericarp is so hard that it can only be split with an axe. In its natural habitat,

opposite: *Ochna natalitia* (Ochnaceae) – coast boxwood; native to southern Africa – immature fruit; in the genus *Ochna* the carpels of the flower are joined only at the base and develop into separate drupelets that sit on an enlarged, fleshy red flower axis

below: *Cyclamen graecum* (Primulaceae) – Greek snowbread; native to Greece, Turkey and Cyprus – close-up of capsule c.1.5cm in diameter. The fruits are strategically placed at ground level where they open with five to seven valves to reveal sticky seeds with a sugary outer layer that attracts ants for dispersal

above: *Nigella damascena* (Ranunculaceae) – love-in-a-mist; native to the Mediterranean – large, inflated capsule containing numerous seeds; formed by 4 to 5 joined carpels (syncarpous gynoecium); the fruit is exceptional in the buttercup family where apocarpous gynoecia of several separate carpels is the rule

opposite: *Nigella damascena* (Ranunculaceae) – love-in-a-mist; native to the Mediterranean – seed; no obvious adaptations to a specific mode of dispersal but displaying a very sophisticated surface pattern. Dispersal is achieved when the seeds are flung out as the capsules sway in the wind; seed 2.6mm long

the Brazilian rainforest, the only animals able to gnaw their way into the seeds are scatter-hoarding agoutis. When, after much hard work with its sharp, chisel-like front teeth the agouti has managed to gnaw a hole into the hard pericarp of the Brazil nut fruit, it is presented with more seeds than it can eat at one sitting. Some of the seeds are therefore hidden and buried underground. It is only from seeds forgotten or left behind by the agoutis that new Brazil nut trees can grow in the wild. Without the help of these animals the seeds of the Brazil nut tree (the familiar Brazil "nuts") would never escape from their fruit.

Fruits in the bed of Procrustes

There are many more capsule-like fruits that simply do not open and so refuse to fit into the botanical definition of a capsule. Bell pepper (*Capsicum annuum*, Solanaceae), cocoa pods (*Theobroma cacao*, Malvaceae) and the ferocious devil's claw (*Harpagophytum procumbens*, Pedaliaceae) are examples of indehiscent capsules. Mankind has long known the nature of this so very human conceptual dilemma. A Greek myth tells the story of a legendary villainous innkeeper called Procrustes who considered his bed the ultimate standard in the world. More unsettling than his narrow mental attitude, however, was a rather gruesome habit he cultivated. He would invite innocent travellers to rest at his house in Eleusis and, once they lay down, tied them to his bed. If his unfortunate guests failed to fit perfectly into the bed, he would stretch or amputate their limbs until they did. It must have come as a great relief to ancient travellers when Theseus forced Procrustes to lie on his own bed and made him fit by cutting off his head and feet, thus defeating him by his own methods.

It would hardly be worth telling the story of Procrustes if the creation of the paradoxical expression "indehiscent capsule" was the only example of its kind. Other oxymoronic terms given to fruits in a Procrustean attempt to provide conformity by violent means are "dry drupe" (for the coconut), "dehiscent drupe" (for the almond) or "dehiscent berry" (for the nutmeg). To accommodate these and many other – especially tropical – types of fruits in a more logical and scientific classification, botanists created a plethora of technical terms and definitions to accompany them, resulting in more than 150 scientific terms for fruits.

False fruits

Just as it is not a simple task to classify fruit morphologically, it is difficult to define scientifically what a fruit is in the first place. In the seventeenth and eighteenth centuries, an era long before the detailed structure of the gynoecium was taken into account, defining a fruit appeared to be a simple task. In 1694 Joseph Pitton de Tournefort (1656-1708) defined a fruit simply as the product of a flower. Less than a century later, Carl von Linné

Illicium verum (Schisandraceae) – star anise; native to Southern China and Vietnam (only known in cultivation) – unripe fruit as it is commercially sold for use as a culinary spice. The fruit is derived from an apocarpous gynoecium and is therefore classified as a multiple fruit; each individual carpel develops into a single-seeded follicle; fruit 4cm in diameter

(1707–1778) in his *Species Plantarum* (1754) and Joseph Gärtner (1732–1791) in his famous book *De fructibus et seminibus plantarum* ("On the fruits and seeds of plants", published 1788–1792) concluded that a fruit is a mature ovary.

Although the simplicity of these definitions may seem both plausible and appealing, they come with a set of rather implausible consequences. For example, compound fruits that develop from more than one flower do not qualify as fruits according to these definitions, and nor do the seed-bearing organs of the gymnosperms since they lack the carpels that constitute an ovary. It is therefore surprising that many botanists still follow the traditional (out-of-date) seventeenth and eighteenth century concepts. They believe that true fruits must be derived from either a single flower or, worse, solely from the gynoecium of a single flower. If any parts of the flower other than the carpels are involved in the formation of the fruit, many textbooks claim – in good Procrustean tradition – that it is a *false fruit* or *pseudocarp*. Being Procrustean, the term *pseudocarp* is, of course, contradictory, in that it refers to a fruit type that is not supposed to be a fruit. In a rather heroic attempt to bring order into the chaos of fruit classification, Richard Spjut (1994) suggested the more appropriate term *anthocarp* for fruits in which attached floral parts persist and develop to form an integral part of the mature fruit (Greek: *anthos* = flower + *karpos* = fruit). Spjut also deserves credit for providing a precise scientific definition of the term "fruit" that for the first time allows botanists to address the seed-bearing organs of the gymnosperms as "proper fruits".

The strawberry effect

Strawberries are the favourite textbook example of a *false fruit* (or *anthocarp*) because most of the edible part is produced by the axis of the flower into which the numerous separate carpels (multiple fruits) are inserted. The individual carpels of a ripe strawberry form single-seeded nutlets, visible as tiny brown granules embedded on the surface of the fruit. What are perceived as hairs or bristles are the remains of the styles, one attached to each nutlet.

Other anthocarps include apples, rose hips and pomegranates (all three with fleshy floral tubes), the pineapple (fleshy inflorescence) and everything that qualifies as a cypsela. Some magnificent examples of anthocarps are produced by the members of the meranti family (Dipterocarpaceae), giant trees that form the dominant component of lowland rainforests in South-East Asia. The five persistent sepals of the flower surround their often large, single-seeded nuts. Depending on the genus, two (*Dipterocarpus, Hopea, Vatica*), three (*Shorea*) or all five (*Dryobalanops*) sepals greatly enlarge during fruit maturation, allowing the wind-dispersed fruits a helicopter-like flight on their protracted journeys to the ground. Since their wings are not formed by the ovary, as in true samaras, they are called *pseudosamaras*.

Bertholletia excelsa (Lecythidaceae) – Brazil nut; native to Brazil – large woody fruit with animal-made hole affording a view of an eroded seed. In its natural habitat, the Brazilian rainforest, the mature fruits fall to the ground where agoutis (cat-size brown rodents) are the only animals able to gnaw their way through the fruit wall into the seeds. They eat some of them and cache the rest for subsequent use. Seeds left in a forgotten cache germinate after 12 to 18 months and grow into new trees; the fruit illustrated is 10cm long

Much of the morphology of a fruit is determined by the inherited structure of the flower, especially the gynoecium. Nevertheless, there are many more kinds of fruits than there are gynoecia. The sheer diversity of fruit types is proof of the enormous flexibility with which angiosperms evolved and diversified. Among all fruits, anthocarps demonstrate best how they developed almost every conceivable modification and combination of organs to achieve their goal: the successful dispersal of their seeds.

The dispersal of fruits and seeds

Unlike animals, plants are rooted in the ground and tied to one place. For most plants, therefore, the seed is the only phase in their life when they are mobile. Travelling as a seed gives a plant the unique chance to escape unwanted competition and other unfavourable conditions and hazards such as predators attracted by the parent plants. In most cases it is not advantageous for a seed to germinate in its place of origin. The young seedling would have to compete for light, water and nutrients, not only with its siblings but also with the mother plant. For this reason, fruits and seeds have often developed special adaptations that allow them to travel. These functional adaptations can be obvious and aesthetic, creating structures that resemble sophisticated pieces of engineering. It is therefore not surprising that the dispersal of fruits and seeds has long fascinated both biologists and the general public.

Depending on the type of fruit, the nature of the dispersal unit (the *diaspore*), varies. In dehiscent (capsular) fruits, which open to release their seeds, it is the seed itself that functions as the diaspore. Fruits or fruitlets that remain closed (berries, drupes, nuts and nutlets) are dispersed with the seed. In tumbleweeds like the Russian thistle (*Salsola kali*, Amaranthaceae) and the tumble pigweed (*Amaranthus caudatus*, Amaranthaceae), the entire plant functions as a diaspore. Irrespective of the nature of their diaspores, plants pursue four principal strategies of dispersal: they can rely on natural processes (wind or water dispersal); have fruits that actively disperse their seeds themselves (self-dispersal); or entice and sometimes also enslave animal couriers into their service (animal dispersal). The enormous diversity of diaspores found among seed plants is predominantly the result of adaptations to these four dispersal mechanisms. Which strategy a diaspore uses, is usually reflected in its "Gestalt" and displayed by a specific syndrome of characters involving colour, texture and size.

Wind-dispersal

In all climates, many plants entrust their diaspores to the wind. Since wind dispersal is such a common practice, the diversity of wind-dispersed diaspores is enormous and an entire volume could easily be devoted to them. Structurally, it is mostly the seeds themselves that

Fruits of the meranti family (Dipterocarpaceae) – native to South-East Asia – depending on the genus, of the five sepals of a flower either two (*Dipterocarpus*), three (*Shorea*) or all five (*Dryobalanops*) enlarge and develop into wings as the fruit matures. Since the wings are not formed by the ovary wall, as in true samaras, the fruits of the Dipterocarpaceae are called pseudosamaras: from left to right, top to bottom: *Dipterocarpus costulatus* (keruing kipas (Malay), 18.5cm long); *Dryobalanops aromatica* (kapur (Malay), 8.5cm long); *Dipterocarpus grandiflorus* (keruing belimbing (Malay), 25cm long); *Dipterocarpus cornutus* (keruing gombang (Malay), 15cm long); *Shorea macrophylla* (engkabang jantong (Malay), 12.5cm long). The members of the meranti family are a dominant component of lowland tropical rainforests and exploited for their valuable timber

The Dispersal of Fruits and Seeds

represent the air-borne dispersal units, less often indehiscent (and then usually single-seeded) fruits. Wind dispersal or *anemochory* has several advantages. Strong air currents or a storm can carry a fruit or seed far away, sometimes many kilometres. Travelling on the wind is also cheap since the energy-rich rewards needed to attract animal dispersers are unnecessary. However, a significant disadvantage of wind dispersal is that the distribution of the diaspores depends on the direction and strength of the wind. Wind dispersal is therefore haphazard and hence wasteful. Most wind-dispersed seeds are doomed because they fail to reach a suitable place where they can grow into a new plant. And so part of the energy that is saved on rewards for more reliable animal dispersers has to be invested in the production of a larger number of seeds. A single capsule of an orchid, for example, can contain up to four million dust-like seeds.

Wind-dispersed diaspores display some distinct functional adaptations. Their structure and shape are adapted to catch as much wind as possible and to increase their buoyancy in the air. This can be achieved through appendages like wings and hairs, by an ultra-light seed coat or pericarp, by the inclusion of air chambers, or by a combination of these adaptations. Whichever organs are involved, the tissues from which these structures are formed usually consist of dead, air-filled cells with thin walls in order to reduce weight and achieve minimum overall density of the diaspore.

Winged diaspores

Wind-dispersed fruits and seeds with wings are common among the gymnosperms and angiosperms. Depending on whether the diaspore is a seed or a fruit, the wing can be formed by the seed coat, the ovary wall, by the enlarged sepals of the flower or subtending leaves (bracts). Wings can be expressed as a single unilateral structure, a pair of opposite blades, a continuous ring surrounding the circumference of the diaspore, or multiple wings. The shape and arrangement of the wings determines the flight characteristics of a diaspore.

Diaspores with a single lateral wing

In maple "seeds" (which are in reality fruitlets of a schizocarpic fruit) and the samaras shed by ash trees the single, one-sided wing permits a helicopter-like flight. During the flight, the diaspore rotates around its centre of gravity, which is located in the thickened seed-bearing end of the fruit. Very similar fruits are found in the legume family (for example, *Tipuana tipu*, *Luetzelburgia auriculata*), the largest of which belong to the Brazilian zebra wood tree (*Centrolobium robustum*). The gigantic wing of this samara – up to 30cm long – is attached to the large spine-covered seed-bearing part. She-oaks, members of the genus

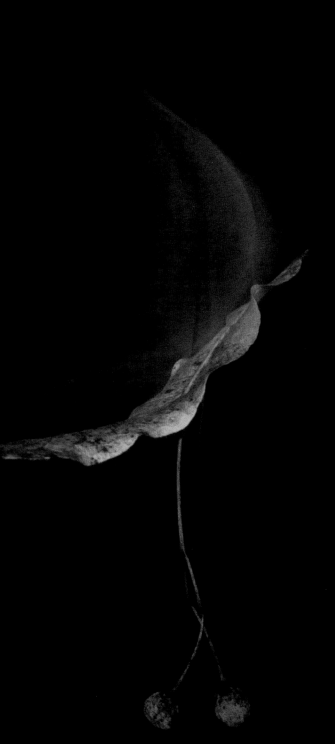

Tilia cordata (Malvaceae) – small-leaved lime tree; native to Europe – the mature (compound) fruit is a cluster of small round nuts covered with grey-brown hair. The nuts are borne in a hanging cluster on the stalk of the inflorescence, which is partly adjoined to a large bract that acts as a wing to helicopter the nuts gently to the ground; the entire fruit is c.8cm long

The Dispersal of Fruits and Seeds

Centrolobium microchaete (Fabaceae) – tarara amarilla; native to South America – fruit (samara) with a single lateral wing attached to the spine-covered seed-bearing part; total length c.15cm. The largest samaras of this type are found in a related species, the Brazilian zebra wood tree (*Centrolobium robustum*) with wings up to 30cm long

Casuarina in the enigmatic she-oak family (Casuarinaceae), also shed their seeds enclosed in samaras with a single wing. Their rather small samaras generally measure less than one centimetre in length and are released from complicated cone-like catkins. These catkins represent compound fruits in which each samara is tightly enclosed by two small leaves (bracteoles), which open only when the fruits are ripe. The lime tree (*Tilia cordata*, Malvaceae) provides a more familiar example of a winged compound fruit. Here the stalk of the inflorescence is partly adjoined to a large bract. Later, the bract acts as a wing to helicopter a small bunch of nuts gently to the ground.

Diaspores of similar shape but representing unilaterally winged seeds rather than fruits are found in both gymnosperms and angiosperms. Among the gymnosperms such seeds are produced by the monkey puzzles (*Araucaria* spp., Araucariaceae) and many members of the pine family (Pinaceae), for example, pines (*Pinus* spp.), spruces (*Picea* spp.), firs (*Abies* spp.), larches (*Larix* spp.) and hemlocks (*Tsuga* spp.). The wing of these seeds is not produced by the seed coat, as with angiosperms, but by a section of the surface of the scale that detaches along with the seed. In many pine species (for example, jack pine, *Pinus banksiana*) the cones are retained on the tree for several years and open to release their winged seeds only after they have been burnt by fire. Such late-opening cones are called *serotinous*.

Within the angiosperms, unilaterally winged seeds have evolved independently in a wide range of families, for example, the bittersweet family (Celastraceae; e.g. *Hippocratea parvifolia*), mahogany family (Meliaceae, e.g. the Cuba mahogani, *Swietenia mahagonii*), mallow family (Malvaceae, e.g. *Pterospermum acerifolium*), Altingiaceae family (e.g. the gum tree, *Liquidambar styraciflua*), and protea family (Proteaceae, e.g. *Banksia*, *Hakea*). Some have rather interesting fruits. The small winged seeds of the gum tree are released from a spherical cluster of small capsules (a compound fruit) curiously resembling a morning star. Members of the Australian genus *Banksia* also release their winged seeds from woody compound fruits but their cones resemble those of gymnosperms and, like the cones of the jack pine, *Banksia* fruits are often serotinous. Well armoured against predators and fire, the tightly closed fruits can remain attached to the plant for years and open only after the parent plants have been destroyed by a field fire. Once they have been exposed to fierce heat the individual fruitlets (follicles) of the *Banksia* "cones" quickly open to release a pair of winged seeds each.

Diaspores with two wings

Diaspores with two opposite horizontal wings either rotate on their longitudinal axis or gently glide through the air on rising thermal currents. The trumpet creeper family (Bignoniaceae) is characterised by wafer-thin, winged seeds. Many have two opposite

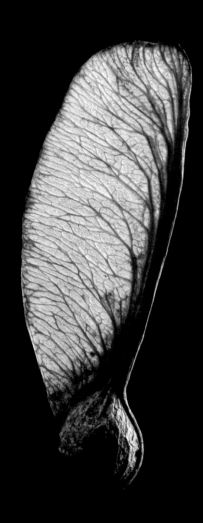

Acer pseudoplatanus (Sapindaceae) – sycamore maple; native to Europe and western Asia – fruitlet; maple fruits develop from two joined carpels that separate into two single-seeded winged nutlets at maturity; the fruitlet is c.5cm long

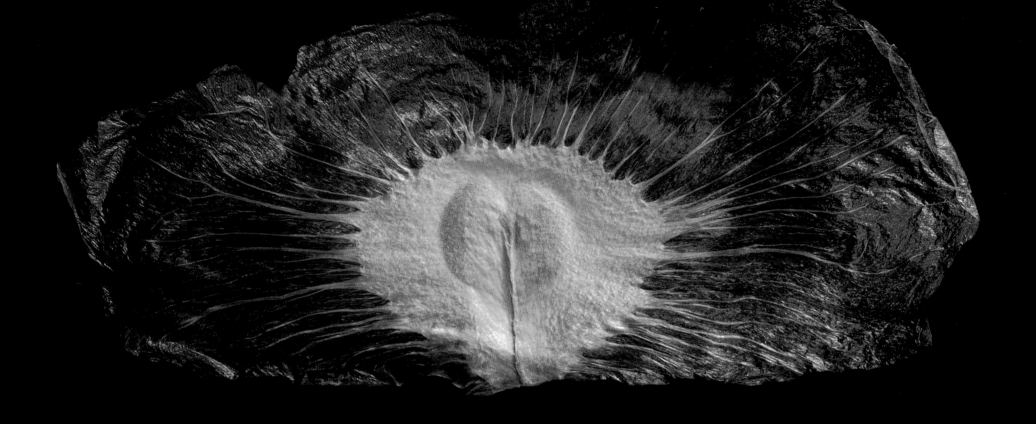

wings, such as those of the African flame tree (*Spathodea campanulata*), the wonga-wonga vine (*Pandorea pandorana*), the monkey pod (*Pithecoctenium crucigerum*) and the yellow trumpet bush (*Tecoma stans*).

As there is considerably less wind in tropical rainforests than in our temperate climates, it is not surprising that the largest wingspan of any gliding seed is found in the jungles of Indonesia. The seeds of the liana *Alsomitra macrocarpa*, a member of the gourd family (Cucurbitaceae), can be 12cm across. They are released from large, pot-like fruits high up in the canopy and in favourable conditions travel for hundreds of metres. Their aerodynamic shape reportedly served early aircraft engineers as a model for wing designs.

Diaspores with more than two wings

Diaspores with multiple apical wings, such as the pseudosamaras of the Dipterocarpaceae family describe a spinning motion similar to diaspores with a single wing. Depending on the force of the wind, fruits with multiple lateral wings follow a more irregular flight path, fluttering or spinning their way to the ground. Some impressive examples of such fruits are found in members of the combretum family (Combretaceae) and mallow family (Malvaceae). The four-winged fruits of *Combretum zeyheri* (Combretaceae), a southern African tree, can reach 8cm in diameter. Similar looking but much larger in size and bearing more wings are the fruits of the malvaceous cuipo tree (*Cavanillesia platanifolia*), a 40m giant from the Central American rainforest.

On a much smaller scale are the tiny winged seeds of certain larkspurs (*Consolida* spp., *Delphinium* spp., Ranunculaceae). A dress of papery lamellae arranged in a spiral around the longitudinal axis of the seeds of *Delphinium peregrinum, D. requienii* and *Consolida orientalis* serves as a device to catch the wind, which lifts them out of their fruits and blows them across the ground beyond the shadow of the mother plant.

Disc-shaped diaspores

The fourth possibility comprises diaspores equipped with a continuous ring-like wing surrounding the central, embryo-bearing part. With the centre of gravity in the middle, such diaspores glide slowly to the ground describing large loops in calm air, but spin or flutter in the wind. Among those producing single-seeded samaras are our familiar elms (*Ulmus*, Ulmaceae), the hop tree (*Ptelea trifoliata*, Rutaceae), which has beautifully patterned samaras, Christ's thorn (*Paliurus spina-christi*, Rhamnaceae), and the wild teak (*Pterocarpus angolensis*), an African tree of the legume family (Fabaceae) with fruits that are not only impressive for their size but also for the long bristles covering the seed-bearing part.

Pithecoctenium crucigerum (Bignoniaceae) – monkey pod or monkey comb; collected in Mexico – wind-dispersed seed with two opposite horizontal wings. The heart-shaped embryo-bearing centre of the seed is marked by the raphe which runs as a raised line from the darker hilum down the middle. According to the regional folklore of Yucatán, Mexico, the Mayan goddess of the jungle Xtabay appears in isolated settlements in the form of a beautiful woman, combing her hair with the rough-surfaced fruits of this liana. Seeds of this shape glide gently through the air on calm days or rotate on their longitudinal axis in the wind; total wingspan os seed 7.6cm

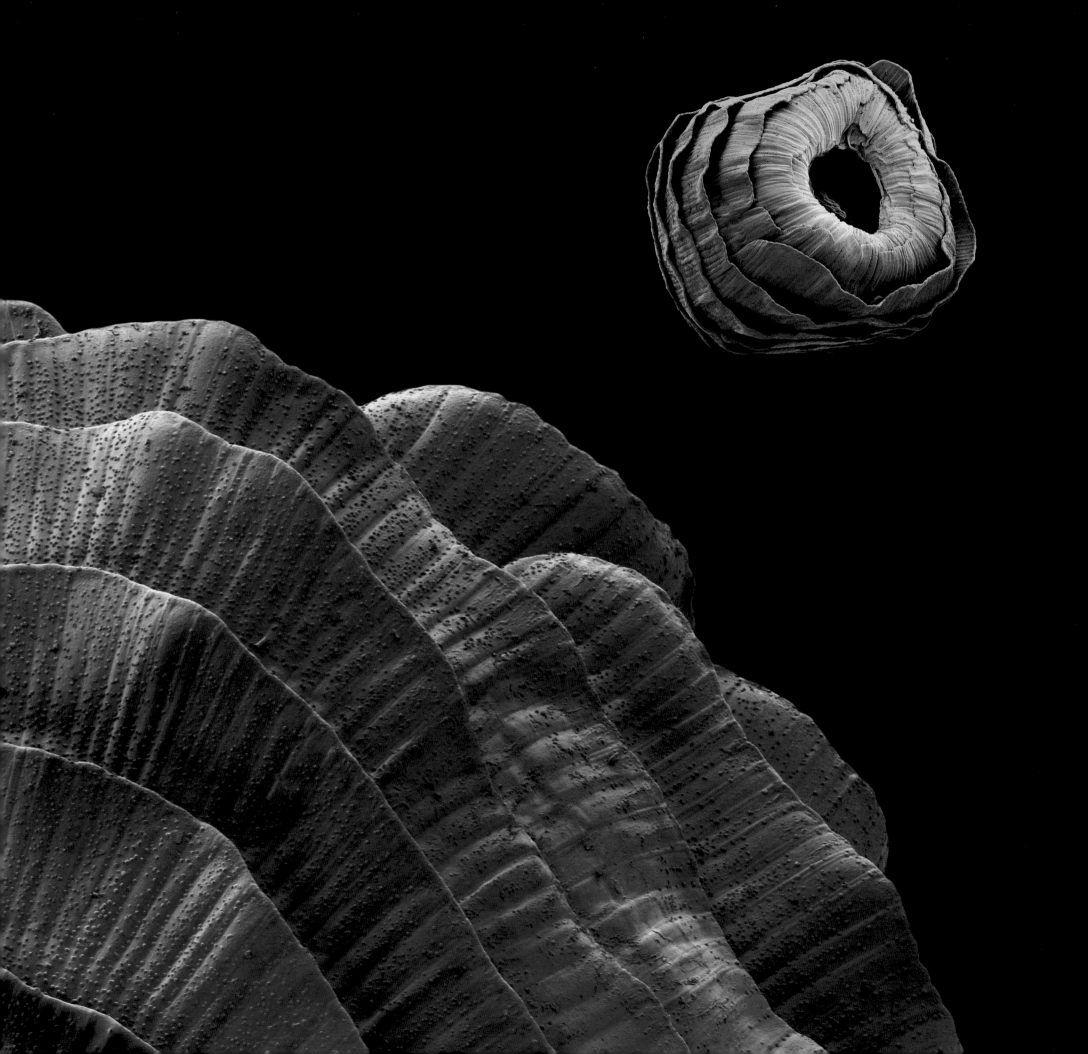

Seeds of larkspurs (Ranunculaceae) wearing a dress of helically arranged, papery lamellae that assist wind dispersal; from left to right: *Delphinium peregrinum* (native to the Mediterranean; 1.3mm in diameter); *Delphinium requienii* (native to southern France, Corsica and Sardinia; 2.6mm long); *Consolida orientalis* (native to southern Europe; 1.8mm in diameter)

Peripheral wings are also found in wind-dispersed seeds, most notably in some members of the Bignoniaceae family. The magnificent jacaranda tree (*Jacaranda mimosifolia*) and the popular ornamental Indian bean tree (*Catalpa bignonioides*) have flat, wafer-thin seeds with a broad wing encircling the embryo-bearing part. Another frequently cultivated tree – similar in aspect to *Catalpa* and also bearing winged seeds that look like those of the Bignoniaceae – is the princess tree (*Paulownia tomentosa*). However, despite the close superficial resemblance to *Catalpa*, anatomical characters place it in its own family (Paulowniaceae) where it is the only tree-like representative among otherwise herbaceous plants. Almost perfectly circular in outline and exhibiting an intricately ornate centre are the paper-thin seeds of the pink velleia (*Velleia rosea*) in the Goodenia family (Goodeniaceae). Miniature examples of similar disc-shaped seeds are provided by the common toadflax (*Linaria vulgaris*, Plantaginaceae) and the greater sea-spurrey (*Spergularia media*, Caryophyllaceae). With a diameter between 1.5mm and 2mm the seeds of the common toadflax are on average slightly larger than those of the greater sea-spurrey, which rarely exceed 1.5mm. Similar in shape and size, but boasting an elaborate surface sculpturing resembling a masterpiece of ultra-light construction are the seeds of the South African *Nemesia versicolor* (Plantaginaceae), a relative of the toadflax.

Hairs, feathers and parachutes

Whereas heavier wind-dispersed fruits are equipped with aerodynamic wings, small diaspores can get by with just a plume of hairs, which affords them sufficient air resistance to float for miles on a slight breeze. There are various possibilities for the arrangement of the hairs (long, air-filled cells) on the diaspore. In the simplest case, the hairs cover the entire diaspore. In some species of morning glory (e.g. *Ipomoea kituiensis*, Convolvulaceae) the seed coat produces long dense hairs. Similar seeds are found in some members of the mallow family (Malvaceae), the most famous being cotton (*Gossypium herbaceum*): the labels on our clothes are proof of the great usefulness of its very long seed hairs. The tiny seeds of poplar (*Populus*) and willow (*Salix*), both in the willow family (Salicaceae), are wrapped in a cloud of hairs produced by the funiculus. These hairs keep the seeds afloat not only in the air but also on water, a double strategy that accords perfectly with their preference for flood plains and other wet habitats. A similar dispersal principle is found in the South American kapok tree (*Ceiba pentandra*, Malvaceae). Inside the large capsules, the smooth, globular seeds are embedded in a mass of white silky hair produced by the carpel walls. These hairs have some valuable properties: their wide air-filled lumina are extremely light and provide good insulation material and stuffing for mattresses. In addition, their outer

Spergularia media (Caryophyllaceae) – greater sea-spurrey; collected in Belgium – seed with peripheral wing that assists wind dispersal; 1.5mm in diameter, including wing

The Dispersal of Fruits and Seeds

layer (cuticle) is waterproof so they never get wet. Kapok can support thirty times its own weight in water, and it is therefore also used to stuff life jackets.

If the hairs are arranged in a more localised manner on the seed, they can appear as one- or two-sided tufts (*comas*) or crowns of hairs. In willowherbs (*Epilobium* spp., Onagraceae) a rather untidy tuft of long hairs adorns the chalazal end of the seeds. The hairs of the seeds in the dogbane family (Apocynaceae) are much more accurately arranged: they form perfectly symmetrical parachutes, either at the micropylar end (*Nerium oleander, Tweedia caerulea*) or at both the micropylar and chalazal ends (as in the desert rose, *Adenium obesum*). When inside the fruit, the stiff hairs of the comas are folded tightly together but as soon as the fruit (a follicle) opens, the parachute unfolds and forces out the seeds in a silky cloud.

Classic examples of parachuted fruits (cypselas) are found in the sunflower family (Asteraceae). Everyone has picked one of the delicate globose "seed heads" of dandelion (*Taraxacum officinale*) or meadow salsify (*Tragopogon pratensis*) and watched the small umbrella-like fruits float on the wind. In the cypselas of the sunflower family the parachute is formed by the modified feathery calyx of the flower and called a *pappus*. The pappus sits at the end of a long, slender stalk, which carries the seed-bearing part of the fruit at the bottom. The slightest gust of wind detaches the fruit from its mother plant and thermals carry it high into the atmosphere where air currents can transport it for many kilometres. Even more elaborate than the delicate cypselas of the Asteraceae are those of the distantly related teasel family (Dipsacaceae). The plumose pappus is replaced by a set of stiff awns. The actual parachute of the fruit is not produced by the flower itself but by the collar of an air bag formed by an outer calyx of four laterally fused bracts surrounding the gynoecium. Overall, the cypselas of the teasel family are heavier and much plumper than their delicate counterparts in the sunflower family. Perhaps that is why they pursue a double strategy: their stiff calyx awns can also easily hook the fruit to the fur of a passing animal.

In the multiple fruits of the buttercup family (Ranunculaceae) the persistent styles of the individual fruitlets sometimes assist wind dispersal if they develop into long, feather-like structures. In traveller's joy (*Clematis vitalba*) the long, hairy styles are responsible for the shaggy appearance of the fruit, which led to the name "old man's beard". The same type of fruit is found in the related pasque flower (*Pulsatilla vulgaris*). Hairs of a special kind and purpose cover the seeds of the eyelash plant, *Blepharis ciliaris* (Acanthaceae). At home in the Mediterranean region, *Blepharis ciliaris* shows some remarkable adaptations to the dry conditions of its habitat. Although the plants die in the dry season, they continue to hold onto their seeds until the next rain arrives. Once the dead remains of *Blepharis ciliaris* are drenched by heavy rainfall, a hygroscopic mechanism triggers the plant not only to expose

Darlingtonia californica (Sarraceniaceae) – Californian pitcher plant; native to the western United States – seed with numerous air-filled projections and a highly hydrophobic surface assisting dispersal by both wind and water; 2.3mm long

Leucochrysum molle (Asteraceae) – hoary sunray; collected in Australia – wind-dispersed fruit (cypsela) with feathery pappus and close-up of the tip of one pappus ray showing an asteraceous pollen grain (perhaps belonging to the same species) caught between the swollen tips of the hairs

opposite: *Tweedia caerulea* (Apocynaceae) – southern star; native to South America – fruit (follicle) with a silky cloud of plumed seeds emerging; each seed carries a perfectly symmetrical parachute of stiff hairs at the front (micropylar) end so that it can travel on the slightest breeze

below: *Clematis vitalba* (Ranunculaceae) – traveller's joy; native to Eurasia and North Africa – close-up of (multiple) fruit composed of numerous single-seeded nutlets (achenes) bearing a long feathery style that allows them to travel on the wind

overleaf: *Epilobium angustifolium* (Onagraceae) – rose-bay willowherb; common in northern temperate regions – left: close-up of capsule pouring out a cloud of hairy seeds; right: seed with a tuft of hairs at the bottom (chalazal) end of the seed assisting wind dispersal; the seed (without hairs) is 0.95mm long

its fruits but also to catapult the seeds several metres away. The seeds are covered with tightly adpressed thick, white hairs, which are transformed as soon as they come into contact with water. Within seconds they become erect and slimy, gluing the seeds to the ground before the wind can blow them away. The size and orientation of the hairs ensure that the root pole of the embryo is positioned as close as possible to the soil surface. The embryo makes the most of the temporarily wet and favourable conditions in the otherwise dry climate of the Mediterranean, swiftly splits the thin seed coat with its cotyledons, and thrusts its root down into the soil within six hours. The same behaviour can be observed in other species of *Blepharis*, such as the klapperbossie (*Blepharis mitrata*) from South Africa.

Anemoballists

A form of indirect wind dispersal is found in plants with dehiscent fruits. The capsules of *anemoballists* are arranged on long, flexible stalks that sway back and forth in the wind. To release their seeds they open small valves (pinks and campions such as *Petrorhagia nanteuilii* and *Silene diocia*) or pores (poppies, bellflowers) in various numbers and arrangements. The swaying of the fruit stalk, which can also be triggered by passing animals and humans, catapults the seeds out of the fruit. To further support their anemochorous qualities, the seeds of the childing pink and carnations (*Petrorhagia nanteuilii*, *Dianthus* spp., Caryophyllaceae) are flattened and slightly concave-convex.

The sacred lotus (*Nelumbo nucifera*, Nelumbonaceae) is an anemoballist of a special kind. With its waterlily-like flowers and aquatic lifestyle this plant has long been considered a close relative of the waterlilies (Nymphaeaceae) but recent research has shown (most surprisingly) that its closest relatives are plane trees (Platanaceae) and members of the Proteaceae family. *Nelumbo* is special for many reasons, including its unique fruit. The centre of the large, waterlily-like flower is occupied by the enlarged floral axis with the shape and appearance of a showerhead. Sunken into this structure are the individual carpels (apocarpous gynoecium), each of which develops into a single-seeded nutlet. The nutlets remain embedded in their compartments in the floral axis until they are ripe, at which point the compartments open wide enough for them to pass through. When the long fruit stalks are shaken by the wind, the nutlets are flung out and thrown into the water where they immediately sink to the bottom. If the stalk of the fruit breaks off, the nutlets may be dispersed over a longer distance. The "shower head" is wider at the top than the bottom and so lands face down on the water. Its spongy air-filled tissue ensures that the fruit floats on the surface, releasing some of the fruitlets instantly, and others later as the remaining air

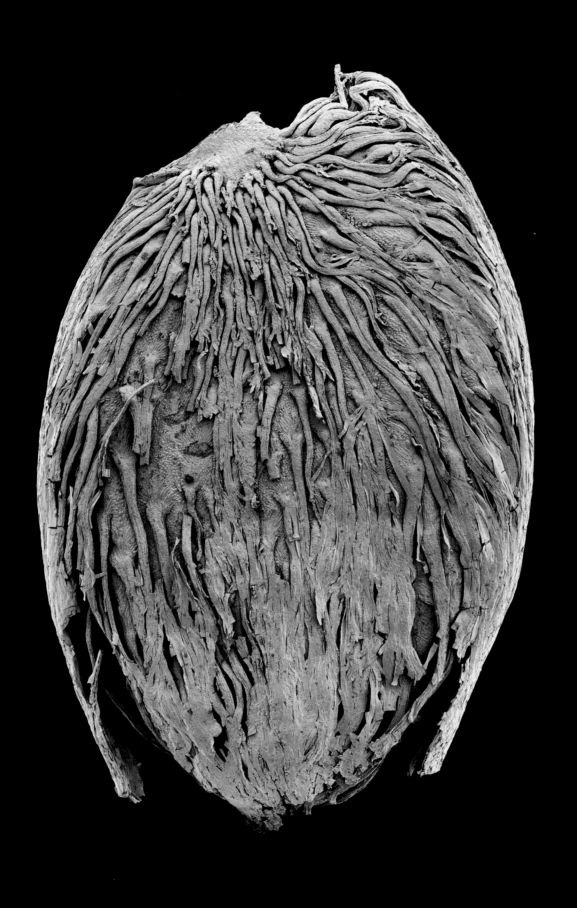

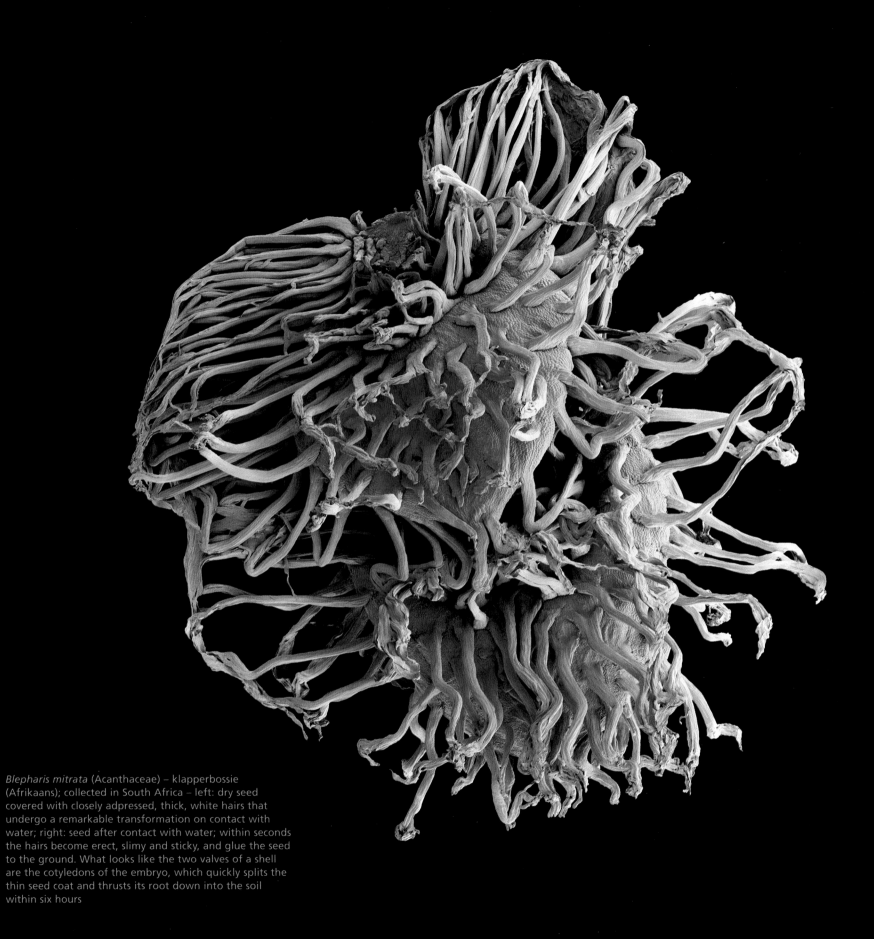

Blepharis mitrata (Acanthaceae) – klapperbossie (Afrikaans); collected in South Africa – left: dry seed covered with closely adpressed, thick, white hairs that undergo a remarkable transformation on contact with water; right: seed after contact with water; within seconds the hairs become erect, slimy and sticky, and glue the seed to the ground. What looks like the two valves of a shell are the cotyledons of the embryo, which quickly splits the thin seed coat and thrusts its root down into the soil within six hours

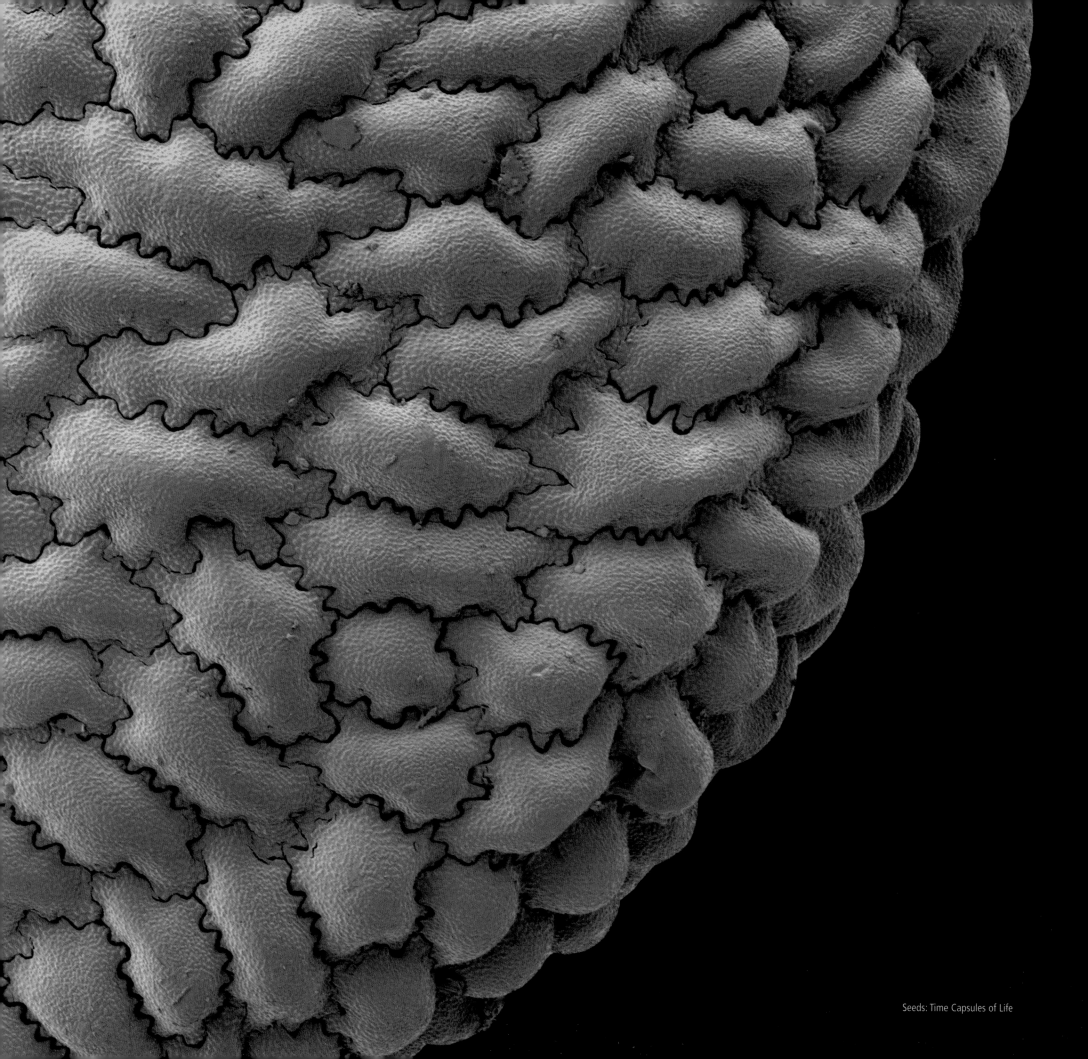

Seeds: Time Capsules of Life

Most anemoballists have more or less spherical seeds without special adaptations, although they do not necessarily lack conspicuous surface sculpturing. Some of the most spectacular surface patterns are found in members of the pink family (Caryophyllaceae), many of which are anemoballists. The cells of their seed coats are interlocked like the pieces of a jigsaw puzzle as shown by the intricate patterns created by their undulating radial walls. Moreover, each testa cell has a convex outer wall. This is often just a smooth bulge (e.g. *Arenaria franklinii*, *Dianthus* spp, *Petrorhagia nanteuilii*, *Silene maritima*) but in many species the testa cells form short papillae (*Agrostemma githago*, *Silene dioica*, *S. vulgaris*) or finger-like projections (e.g. *Stellaria holostea*) that give the seeds a spiny appearance. Interestingly, the ragged robin (*Lychnis flos-cuculi*) has spiny and smooth seeds side by side in the same fruit.

Balloon fruits

An alternative to the attachment of aerodynamic wings is to increase the surface of the diaspore and at the same time reduce its specific weight. This can be achieved through an abundance of hairs, as in cotton seeds, or the inclusion of large air spaces. In fruits, various organs can form such air chambers. A common solution is an inflated ovary creating a bladder or balloon-like fruit with a very thin, often transparent wall. Fruits of this type occur in some well-known ornamental plants such as the lesser honeyflower (*Melianthus minor*, Melianthaceae), the bladder senna (*Colutea arborescens*, Fabaceae), the balloon pea (*Sutherlandia frutescens*, Fabaceae), the goldenrain tree (*Koelreuteria paniculata*, Sapindaceae), and American bladdernut (*Staphylea trifolia*, Staphyleaceae). The principle of including air spaces to achieve lightweight structures with large surfaces has also been adopted by seeds, as will be discussed in the next paragraph.

Dust and balloon seeds

The most effective strategy for ensuring long-distance dispersal by the wind is the production of a large number of extremely small, lightweight seeds. To give some idea of the dimensions involved, a single capsule of the tropical American orchid *Cycnoches chlorchilon* produces almost four million seeds, and one gram of the smallest wind-dispersed orchid seeds (e.g. *Calanthe vestita*) contains more than two million of them. These "dust seeds", hardly larger than the ovules from which they developed, achieve weight reduction at the expense of endosperm and embryo. On dispersal, the amount of endosperm – if present at all – is often negligible and the tiny embryo is merely a globular lump of 8-150 undifferentiated cells. It has been reported that the embryo of the yellow bird's-nest (*Monotropa hypopitys*, Ericaceae) consists of three cells surrounded by nine endosperm cells.

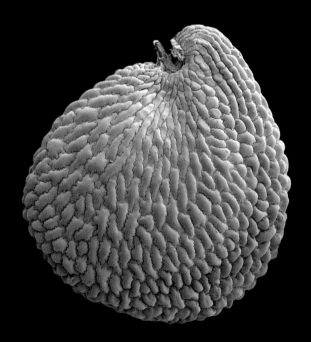

below: *Arenaria franklinii* (Caryophyllaceae) – Franklin's sandword; collected in Idaho, USA – seed with no obvious adaptations to a particular mode of dispersal but displaying the intricate surface pattern typical of the pink family with the cells of the seed coat interlocked like the pieces of a jigsaw puzzle; seed 1.3mm in diameter

opposite: detail of seed coat

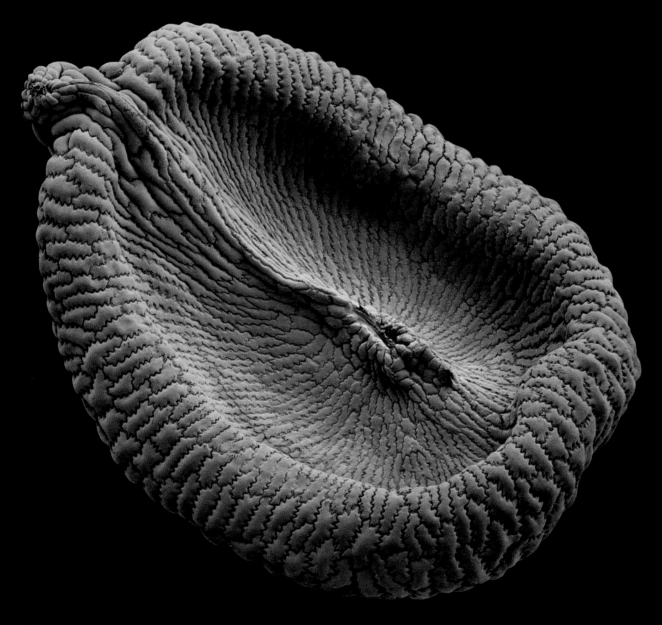

Petrorhagia nanteuilii (Caryophyllaceae) – childing pink, collected in the UK – left: seed, without obvious adaptations to a particular mode of dispersal but displaying the intricate surface pattern typical of the pink family; the flattened, convex shape of the seed (large surface/volume ratio) may assist wind dispersal; 1.2mm long

opposite: close-up of inflorescence

The Dispersal of Fruits and Seeds

The large surface/volume ratio of such "dust seeds" significantly reduces their sink velocity in the air: a small orchid seed sinks at about 4cm per second, which is slow compared to the samara of an elm, which falls at a speed of 67cm per second. Air buoyancy is further increased by a syndrome of obvious adaptations such as air pockets, which may consist of large, empty cells, intercellular spaces, or a space between the seed coat and the embryo-bearing centre of the seed. Seeds equipped with such air spaces are commonly called "balloon seeds". The difference between dust seeds without air pockets and balloon seeds is merely adaptational and both models can be found within the same family (e.g. Droseraceae). Typical dust seeds without air chambers are those of the broomrapes (*Orobanche* spp., Orobanchaceae), begonias (*Begonia* spp., Begoniaceae) and many members of the Ericaceae family (e.g. *Erica* spp., *Rhododendron* spp.). Similar seeds are found in certain species of sundew (e.g. *Drosera cistiflora*, *D. uniflora*, *D. intermedia*, Droseraceae), in butterworts (*Pinguicula* spp., Lentibulariaceae), some members of the gentian family (e.g. *Blackstonia perfoliata*, *Wahlenbergia hederacea*, Gentianaceae) and saxifrage family (e.g. *Heuchera* spp., *Saxifraga* spp.), and Podostemaceae (*Weddellina squamulosa*). Balloon seeds are most famously found in orchids but also in the following families: Droseraceae, including a number of sundew species (e.g. *Drosera anglica, D. natalensis, D. rotundifolia*); Parnassiaceae such as the fringed grass of Parnassus (*Parnassia fimbriata*); Gentianaceae such as marsh gentian (*Gentiana pneumonanthe*) and one-flowered fringed gentian (*Gentianopsis simplex*); Ericaceae such as one-sided pyrola (*Pyrola secunda*); foxgloves (*Digitalis* spp., Plantaginaceae), and many Orobanchaceae such as Indian paintbrushes (*Castilleja* spp.), louseworts (*Pedicularis* spp.), owl's clovers (*Orthocarpus* ssp.), and members of the genera *Lamourouxia* and *Cistanche*.

To further increase the surface-volume ratio, many balloon seeds adopt a narrow tube- or spindle-shape by extending the seed coat at opposite ends, i.e., at the micropylar and chalazal pole. This strategy is taken to the extreme in the long, almost thread-like seeds of the bog asphodel (*Narthecium ossifragum*, Melanthiaceae, which has seeds up to 8mm long), the genus *Pitcairnia* (Bromeliaceae, seeds up to 18mm long) and certain tropical orchids (*Acanthephippium papuanum*, *Sobralia macrantha*). Remarkable exceptions among the almost uniformly spindle-shaped seeds of the orchid family are found in *Cyrtosia*, *Eulophia* and *Stanhopea*: all three possess flattened, wing-like seeds. Completely unexciting in comparison are the seeds of the economically important orchid, the vanilla vine (*Vanilla planifolia*). Rather than relying on the wind for their dispersal, vanilla pods are eaten by bats and the seeds germinate after passing through their gut. This explains the absence of the balloon seed syndrome in vanilla seeds, whose smooth black discs are responsible for the tiny dark spots in vanilla ice-cream.

Drosera cistiflora (Droseraceae) – cistus-flowered sundew; native to South Africa (Western Cape) – seed displaying the typical honeycomb pattern of wind-dispersed dust seeds; the "freak" cell in the middle repeats the pattern of the seed coat; the soft, velvety texture of the surface is created by a layer of wax granules; 0.5mm long

overleaf left: *Narthecium ossifragum* (Melanthiaceae) – bog asphodel; collected in the UK – extremely elongated, almost thread-like wind-dispersed balloon seed; c.6mm long

overleaf right: close-up of flowers; the bog asphodel is found on damp heaths and peatbogs. Its Latin name *ossifragum* means "bone-breaker" and refers to the ancient belief that the bones of any sheep that ate it would become brittle

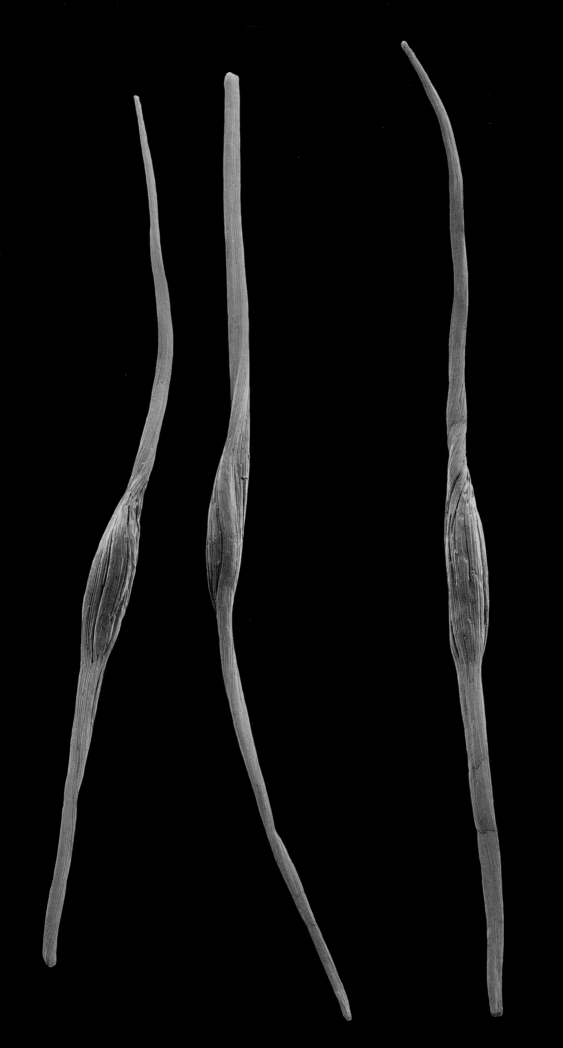

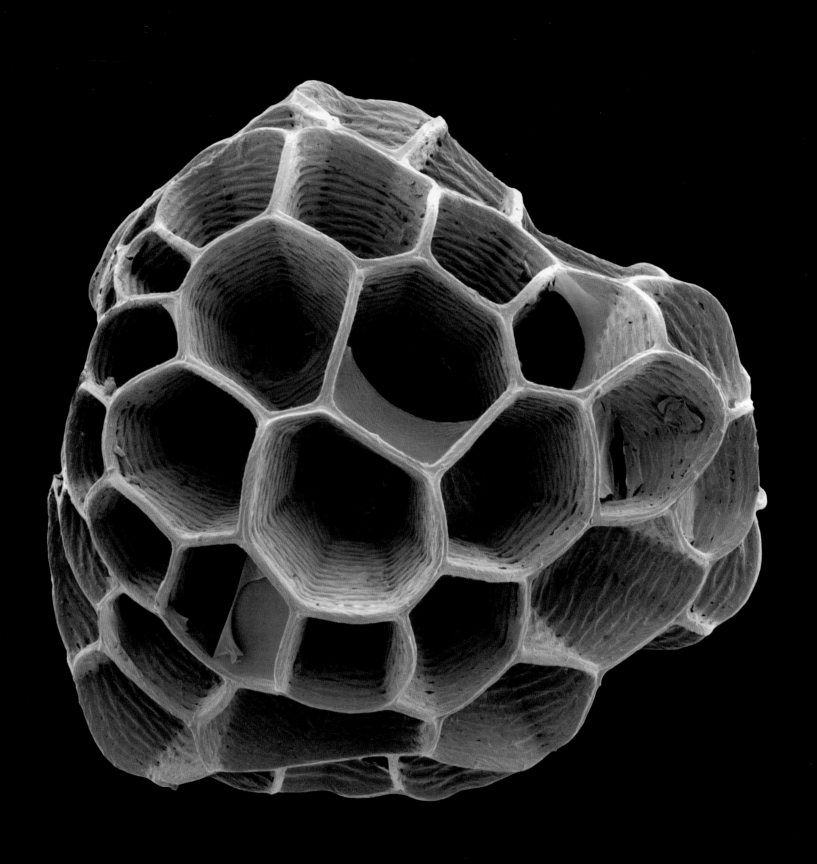

The honeycomb principle

Although largely unrelated, families with the dust and balloon seed syndromes display a striking convergence. In many if not most, the single-layered seed coat has a distinct honeycomb pattern with either isodiametric or elongated facets. The honeycomb pattern ensures maximum stability for the structure with minimum thickness and thus minimum weight for its load-carrying parts. This is a universal principle and honeycomb patterns can be observed in both the inanimate and animate world. They appear in the arrangement of carbon atoms in graphite; bees use them to construct their storage containers for honey; they are reflected in the surface structure of certain pollen grains; and in modern engineering honeycomb cores confer high stability on sandwich structures (for example, doors, lightweight components for aeroplanes). In the case of dust and balloon seeds, the honeycomb pattern is produced by the dead, air-filled cells of their single-layered seed coat. Their radial walls are slightly thickened whereas the outer tangential walls remain thin, and collapse as the cells dry out. This not only reveals the internal honeycomb pattern of the seed coat but also increases the surface area of the seed. In most balloon seeds, the outer and inner tangential walls remain intact. However, there are some interesting variations. In seeds of the foxglove (*Digitalis* spp.) and the related genus *Anarrhinum* (both members of the plantain family, Plantaginaceae), the outer tangential wall has totally disappeared at maturity, exposing the radial walls forming the honeycomb pattern. The loss of the outer wall helps reduce the weight of the seed and by exposing the network of radial walls further enlarges its surface area. The strategy of saving material and weight is even expressed in the astonishingly regular microstructure of the radial walls: vertical series of unenforced circular or oblong areas create a grid-like relief which – in engineering terms – is again merely a variation of the material-efficient honeycomb principle. In some members of the broomrape family (Orobanchaceae) like the bird's beak (*Cordylanthus* spp.) and some species of owl's clover (e.g. *Orthocarpus luteus*), the outer tangential wall of the seed coat also disappears. What becomes visible underneath is the complicated reticulate pattern of the inner tangential wall, which seems to add to the stability of the seed coat. The third and most impressive variation of the honeycomb-patterned seed coat is also found in the broomrape family (e.g. *Castilleja*, *Cistanche* spp., some *Orthocarpus* spp.). Here both the outer and inner tangential walls dissolve to leave just the loose, voluminous honeycomb cage produced by the thickened radial walls. A possible explanation as to why these seeds totally expose their reticulate cell wall system in this way is provided by an interesting observation about some owl's clovers (*Orthocarpus* spp.). Owl's clovers are root parasites, especially on members of the sunflower family (Asteraceae). Within the genus *Orthocarpus*, three seed-coat

Cistanche tubulosa (Orobanchaceae) – collected in Saudi Arabia – balloon seed displaying the most extreme form of the typical honeycomb pattern found in wind-dispersed seeds: both the outer and inner tangential walls of the seed coat cells dissolve leaving the loose, voluminous honeycomb cage produced by the thickened radial walls

types are found but only one displays the extreme honeycomb cage pattern. It has been reported that seeds of the latter type tend to become caught in the pappus bristles of the fruit (cypsela) of their host plant, the smooth catsear (*Hypochaeris glabra*, Asteraceae), which often occurs together with *Orthocarpus*. This astonishing joint dispersal (by the wind) of parasite and host ensures that the owl's clover germinates close to its future host plant.

A remarkable convergence

The long list of examples where dust and balloon seeds can be found shows that they occur in a wide variety of largely unrelated families across the angiosperms. That their similarities are due to evolutionary convergence rather the result of a shared ancestry is proven by the fact that the same shapes and patterns occur in distantly related groups. The seeds of some *Pinguicula* species (Lentibulariaceae), for example, are very similar to those of some *Drosera* species in the unrelated sundew family (Droseraceae). An even more astonishing example is *Pyrola secunda* in the heather family (Ericaceae), which has seeds that are indistinguishable in appearance from those of some orchids.

When comparing the lifestyles of plants with dust or balloon seeds, it is striking that most of them belong to families that have to solve some kind of nutritional problem. These are either epiphytes, insectivors and other bog plants, partial or complete parasites, or chlorophyll-less mycotrophic plants. Epiphytes live on branches high up in the canopy where little water and few nutrients are available (e.g. epiphytic orchids). Plants living in acidic and nutrient-poor soils such as bogs face the same problem (e.g. Parnassiaceae). Some bog plants have become insectivors to top up their nitrogen from animal proteins (e.g. Droseraceae, Lentibulariaceae). One of the most bizarre and interesting groups of angiosperms are chlorophyll-less mycotrophic plants. They rely entirely on fungi for their nutrition and so have little, if any, chlorophyll to allow them to assimilate their own sugars. Among them are members of the Ericaceae, like the yellow bird's-nest (*Monotropa hypopitys*), wintergreens (*Chimaphila* spp., *Pyrola* spp.) and Indian pipe (*Monotropa uniflora*), the enigmatic Burmanniaceae family, and some orchids such as the bird's-nest orchid (*Neottia nidus-avis*) and the coral-root orchid (*Corallorhiza trifida*). Partial or hemi-parasites such as the Orobanchaceae (e.g. *Castilleja, Cordylanthus, Lamourouxia, Orthocarpus, Pedicularis*) are root parasites. Although they are green and therefore able to photosynthesize, a substantial portion of their carbohydrates are derived from the host plant through its roots. Some hemiparasites can live on their own but perform better if they find a convenient host.

Why all these different nutritional specialists are connected by the dust and balloon seed syndrome is unclear. Mycotrophic plants and root parasites face the challenge that their

Orthocarpus luteus (Orobanchaceae) – yellow owl's clover; collected in Oregon, USA – balloon seed displaying the typical honeycomb pattern of wind-dispersed seeds. The outer tangential wall of the cells of the seed coat disappears, revealing the intricate pattern of the inner tangential walls (coloured brown); seed 1.3mm long

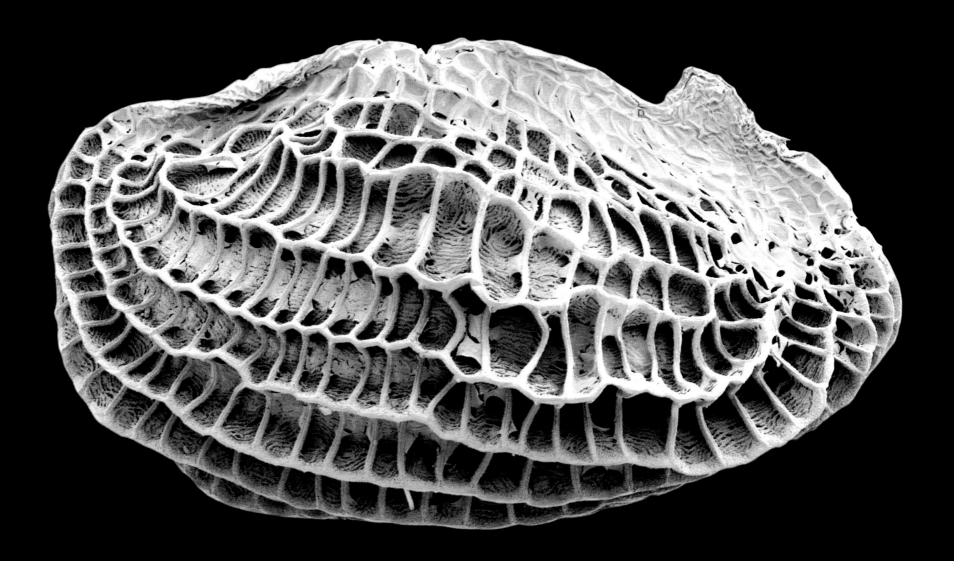

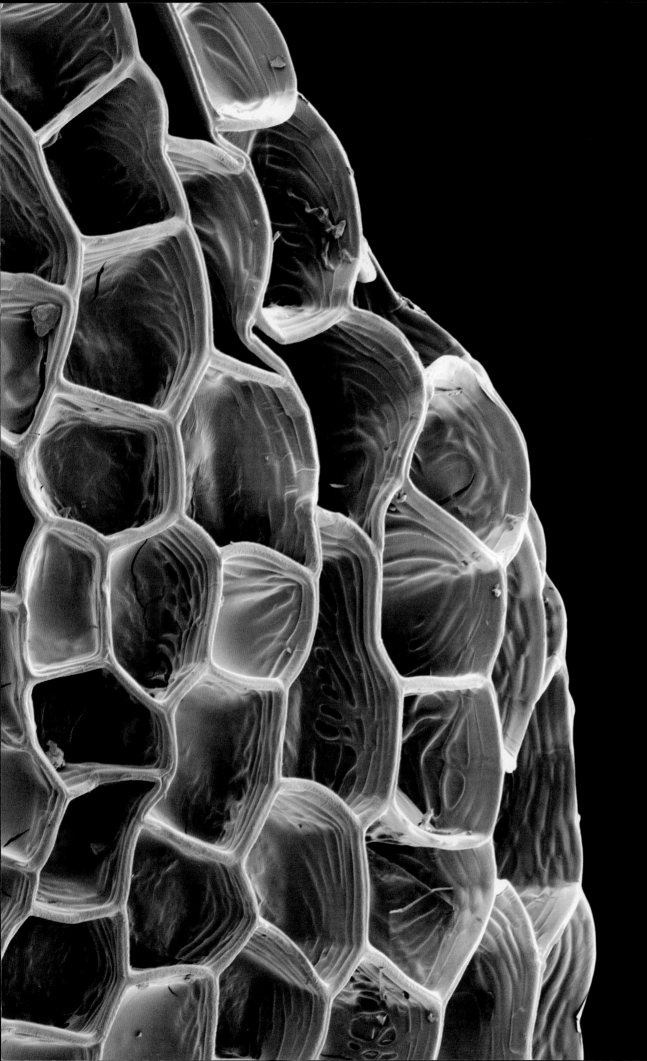
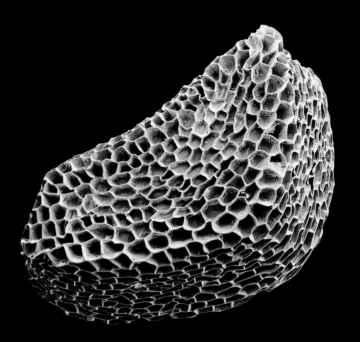

left and above: *Digitalis lutea* (Plantaginaceae) – yellow foxglove; collected in Belgium – seed and detail of seed coat displaying the typical honeycomb pattern of wind-dispersed balloon seeds. In foxglove seeds the outer tangential wall disappears at maturity, exposing the radial walls that form the honeycomb. The strategy of saving material and weight is contained in the vertical series of unenforced cell wall areas in the microstructure of the radial walls; 1.3mm long

opposite: *Digitalis ferruginea* (Plantaginaceae) – rusty foxglove; native from southern Europe to western Asia – close-up of inflorescence

seeds have to meet at least one suitable host (fungus or plant) in order to germinate successfully. The same applies to most orchids, which are unable to germinate without engaging in a mycorrhizal relationship with their specific fungi. The strategy behind the dust and balloon seed syndrome is therefore not primarily long-distance dispersal. It merely ensures a certain statistical probability that at least some of the many seeds shed will end up in a location that fulfils their extremely specific germination requirements. Long-distance dispersal means that the same amount of seed is distributed over a larger area, which could actually lower the odds of encountering a compatible host in a suitable location. The fact that many of the species involved are endemics with very limited distribution supports this theory. This does not mean, however, that dust and balloon seeds are unable to travel long distances. Orchids, for example, manage to reach isolated islands far from the mainland. As famously documented, they were among the first pioneers to resettle on the islets of Krakatoa after the catastrophic volcanic eruption of 27 August 1883.

The clear link between the structure and function of the dust and balloon seed syndrome has already been pointed out. Apart from achieving a high surface to weight ratio, dust and balloon seeds have other interesting physical properties. Firstly, their surface is almost unwettable. This is due both to the chemistry of the seed coat and its reticulate pattern. Many dust and balloon seeds can drift on the water surface without ever getting wet. When they reach a solid substrate, rain washes them into the finest cracks in the bark of trees or in the ground. At this point, their rough surface sculpture helps anchor them onto the substrate. Soluble humic substances present on bark or in the soil then cause changes in the seed coat, rendering it wettable so that it soaks up the water needed for germination.

Water dispersal

The possibility that water assists in the dispersal of fruits and seeds has already been mentioned in passing. Wind-dispersed diaspores such as plumed fruits and seeds (e.g. of willows and poplars), balloon fruits (e.g. of hornbeams) and balloon seeds can often be transported by water also, even if this happens accidentally. Winged diaspores, if small enough, can also stay afloat thanks to the surface tension of the water; the tiny winged seeds of *Spergularia media*, for example, are able to float for many days. The dispersal of primarily wind-dispersed diaspores by water is hardly surprising since the physical strategy behind their adaptations, a low surface/volume ratio, not only affords them floatability in air but often (automatically) in water also. The diaspores of some specialists, particularly aquatic plants, marsh and bog plants and others living in close proximity to water, do, however, possess obvious, specific adaptations to water dispersal. The most important quality of water-

opposite: *Pyrola secunda* (Ericaceae) – one-sided pyrola; collected in the UK – seeds with loose, bag-like seed coat displaying the typical honeycomb pattern of wind-dispersed balloon seeds; c.0.5mm long; the seeds of this species are hardly distinguishable from those of certain orchids (see overleaf)

overleaf left: *Dactylorhiza fuchsii* (Orchidaceae) – common spotted orchid; native to Europe – inflorescence; this species is the most common orchid in the UK and is found in much of Europe except the south

overleaf right: *Dactylorhiza fuchsii* (Orchidaceae) – seed with loose, bag-like seed coat displaying the typical honeycomb pattern of wind-dispersed balloon seeds; c.0.6-0.7mm long

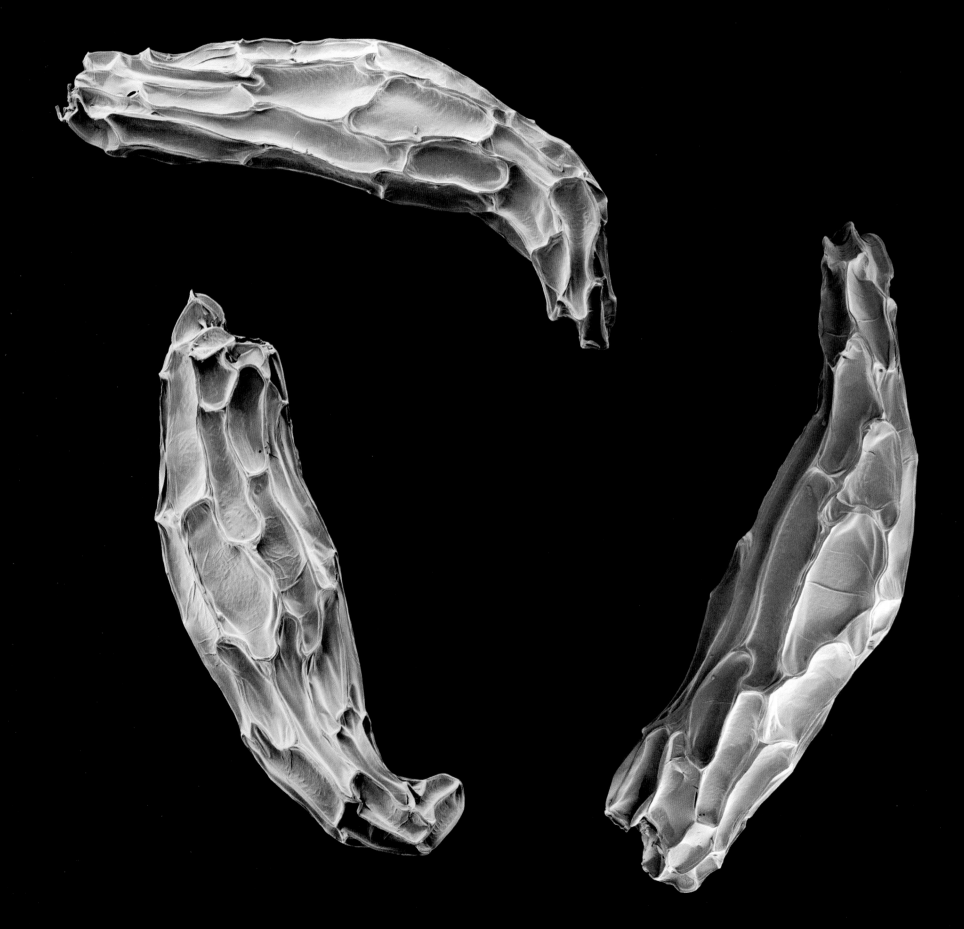

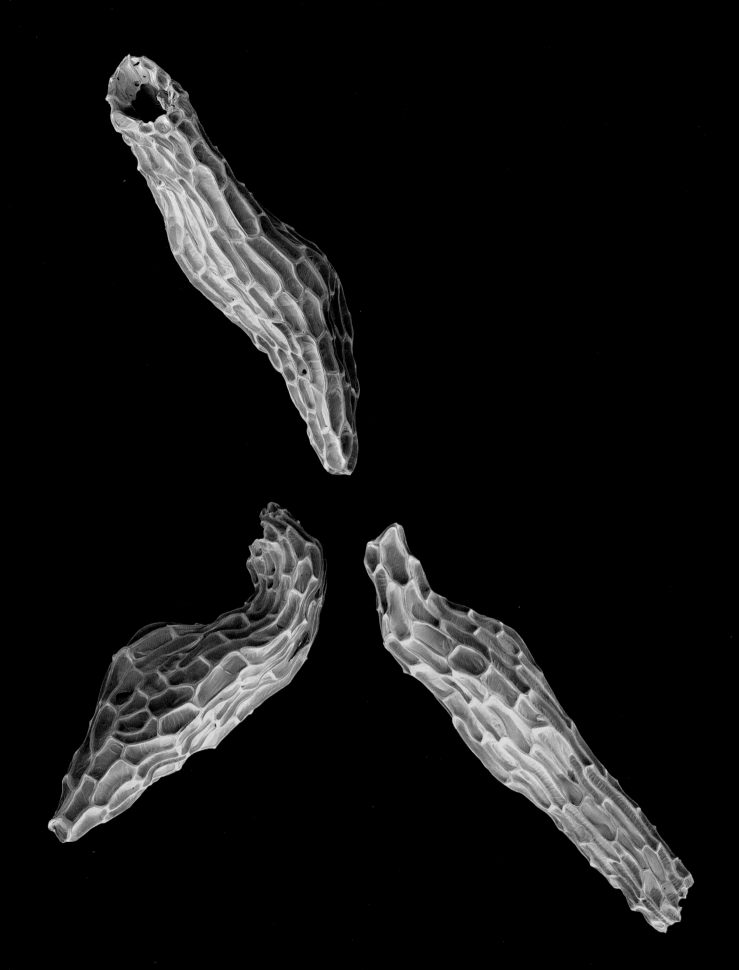

The Dispersal of Fruits and Seeds

dispersed diaspores remains their floatability, which is often enhanced by the addition of water-repellent tissues, especially on the outside. Recognisable buoyancy devices in the form of air spaces and waterproof corky tissues are common. The seeds of the giant waterlily (*Victoria amazonica*) and other Nymphaeaceae (e.g. *Nymphaea alba*) are surrounded by a silvery-translucent airbag of funicular origin; its soft, short-lived tissue keeps the seeds afloat for just a few days. Cork tissues consist of dead, air-filled cells whose walls are impregnated with suberin, the chemical substance that gives cork its waterproof qualities. A massive corky mesocarp with large intercellular air-spaces affords the fruits of *Cerbera manghas* (Apocynaceae) excellent, long-lasting buoyancy. At home in the mangrove swamps along the edge of the Indian Ocean and western Pacific, this tree owes its wide distribution to the seaworthiness of its fruits.

The hooks or spines often present on water-dispersed diaspores help them to anchor on a suitable substrate and cling to the fur or feathers of animals or birds. The seeds of the aquatic yellow floatingheart (*Nymphoides peltata*, Menyanthaceae) combine several of these adaptations. Once the fleshy parts of the fruit have rotted away or been eaten by snails, they open at the base to release the seeds directly into the water. Their flat, discoid shape, the fringe of stiff hairs around the periphery and their water-repellent surface allow them to use the surface tension of water to avoid sinking. Although heavier than water, the seeds can float for two months if not disturbed. Their bristly hairs enable them to form little chains or rafts on the surface of the water, or to hitch a ride on a water bird. Probably the most ferocious hooks of any water-dispersed diaspore adorn the fruits of the water chestnuts (*Trapa* spp., Lythraceae). Their two or four sharp, curved spines not only cling to large birds and other animals or anchor the fruits to the ground, but can also cause humans painful injuries when stepped on. Inside each water chestnut is a single large starchy seed, which is edible. One species, *Trapa natans* (water caltrop, horn nut), has been cultivated in China for more than three thousand years but the crunchy water chestnuts commonly used in Chinese cooking are the fleshy corms (not seeds) of the totally unrelated spike rush (*Eleocharis dulcis*, Cyperaceae).

The diaspores of other aquatic plants (e.g. hornwort, *Ceratophyllum* and water-starword, *Callitriche*), especially those growing in deeper water, often lack buoyancy aids. Their seeds need to germinate on the submerged ground of their flooded habitat and often sink instantly but currents may drive them some distance over the ground. Another possibility is that the dispersal phase of non-floatable diaspores is delayed until after germination. Seeds of the purple loosestrife (*Lythrum salicaria*, Lythraceae) and the flowering rush (*Butomus umbellatus*, Butomaceae) germinate where they sink but their seedlings then rise to the surface, to be dispersed by currents or the wind.

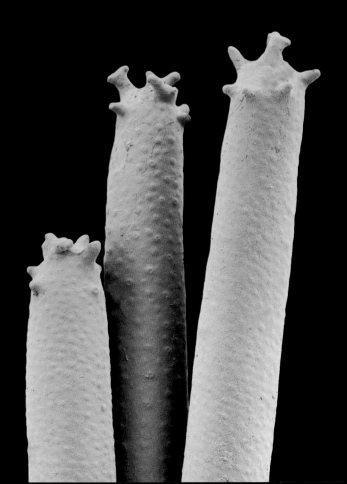

Nymphoides peltata (Menyanthaceae) – yellow floatingheart; collected in the UK – water-dispersed seed and close-up of marginal hairs. Although heavier than water, the flat, discoid shape, water-repellent surface and fringe of stiff hairs allow the seed to use surface tension to avoid sinking. The bristly hairs around the periphery also stick to the feathers of water birds; seed (including bristles) c.5mm long

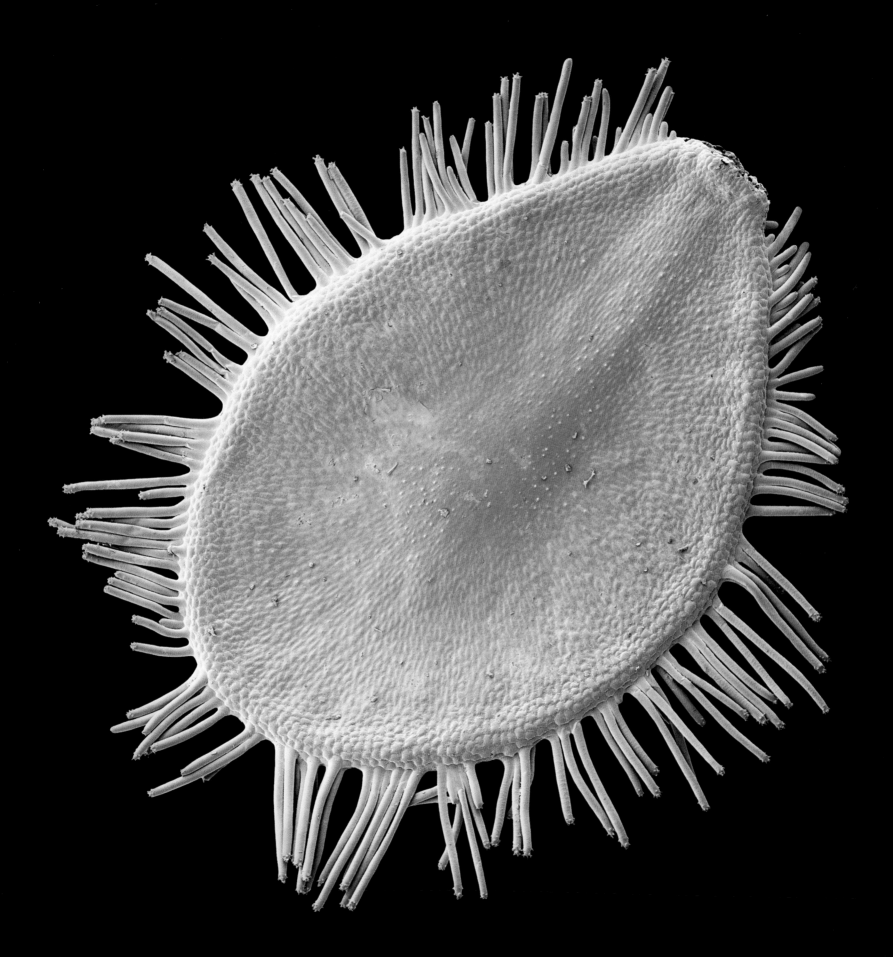
The Dispersal of Fruits and Seeds

Dispersal by rain

Water in the form of rain offers other interesting dispersal opportunities. Rare specialists, such as the members of the stone plant family (Aizoaceae), rely on raindrops to flush their seeds out of their capsules. Although the seeds are not adapted to float, they are small enough to be splashed out. Most Aizoaceae, such as living stones (*Lithops* spp.) and their relatives, live in the dry regions of southern Africa where water is not always available. Whereas most capsular fruits open as they dry out, the opening mechanism of the capsules of the Aizoaceae is triggered only when the fruits get wet. In addition, the *hygrochastic* movement of the capsules' valves is reversible, in other words they can open and close repeatedly. This ensures that their seeds are dispersed only when the water necessary for germination is freely available. A more common sight in temperate latitudes are the fruits of the marsh marigold (*Caltha palustris*). A member of the buttercup family (Ranunculaceae), it has multiple fruits consisting of 9-12 individual fruitlets (follicles). When these follicles open along their upper side to present their seeds, they form a bowl ready to catch any falling drops of rain or dew. When the drops hit the follicles, the seeds are splashed out. Since the marsh marigold usually grows in shallow water at the edges of ponds or streams, the seeds are bound to land in water, where they float. The air-filled tissue at their chalazal end enables them to stay afloat for up to four weeks.

In a similar way to anemoballists, rain-ballists have developed elastic structures to catapult their diaspores away from the plant. The kinetic energy needed to accelerate their diaspores is absorbed from falling rain drops via a springboard mechanism. In members of the mint family (Lamiaceae) such as the skullcap (*Scutellaria galericulata*) and selfheal (*Prunella vulgaris*), the springboard is formed by the calyx of the flower. Rain drops hitting the flexible spoon-shaped upper lip of the calyx bend the flexible stalk of the flower, which rebounds to eject the nutlets. Some genera of the mustard family (Brassicaceae) with flattened fruits such as the field pennycress (*Thlaspi arvensis*) have a different strategy. When their fruits are hit by a drop of water the elastic stalk bends down and bounces back, flinging out the two papery sides of the fruit with the seeds inside. A strong wind can achieve the same result.

Mangroves

Mangroves are tropical and subtropical trees and shrubs most commonly found in the tidal zones of the oceans. The name refers to a life form rather than a group of related plants. Worldwide, more than fifty species in sixteen different families are deemed to be mangroves. Having adapted to conditions that exclude most other plants, they grow in muddy swamplands regularly inundated by the tides. Apart from their characteristic stilt roots, the

Caltha palustris (Ranunculaceae) – marsh marigold; native to northern temperate regions – mature (multiple) fruit prior to opening and fruit with open follicles ready to catch any falling drops of rain or dew that will splash or bounce out the seeds. Growing in shallow water at the edges of ponds and streams, the marsh marigold has seeds with a buoyancy device at the bottom (chalazal) end in the form of a white (later brown) air-filled tissue, which enables them to float for up to four weeks

most significant morphological adaptation to their extreme habitat is their method of reproduction. Rather than shedding ordinary seeds, most of which would be washed away by the tides and lost, mangroves are viviparous. Vivipary (live birth) in animals means that the embryo grows inside the mother (as in most mammals) not inside an egg (as in most reptiles and birds). Viviparous plants such as mangroves produce seeds that germinate while still attached to the parent plant. Once the zygote has been fertilised the embryo simply continues to grow. By extending its hypocotyl, it soon penetrates the thin seed coat and breaks through the wall of the berry. A fully-grown embryo of the most common red mangrove (*Rhizophora mangle*, Rhizophoraceae) may be 25cm long. Eventually, the club-shaped "seedling" drops off and either plants itself immediately in the mud underneath or floats in the sea until the next tide sets it down elsewhere. The advantages of vivipary in the mangroves' habitat are clear: by supporting their embryos until they have grown into sizeable, well-differentiated plantlets, mangroves give their offspring a head start. Ready to go, and equipped with some reserves, mangrove seedlings root very quickly once they touch ground.

Ocean travellers

The seeds and fruits of many plants growing on or in the vicinity of coastlines eventually end up in the sea, where they are carried away by ocean currents. Fruits and seeds may be shed directly on the beach or drop into tidal pools and swamps from where the tide collects them. Those originating further inland reach the sea via streams and rivers, in many cases accidentally. However, a number of plants, especially in the tropics, possess diaspores specifically adapted to travel in sea water for months or even years; one example, the fruit of the mangrove tree *Cerbera manghas*, has already been mentioned. Once such seaworthy diaspores reach the main ocean surface currents, their journey can take them thousands of kilometres away from their place of origin. Charles Darwin was enthralled by the idea that seeds from tropical countries could travel to Europe. Some people collect exotic drift seeds and fruits, popularly known as "sea beans", as a hobby.

Fascinating though it may be, drifting with ocean currents is – like wind dispersal – a very haphazardous and wasteful strategy. Many drift fruits and seeds lose their buoyancy and are likely to end up on the bottom of the ocean or somewhere with unsuitable living conditions. For example, tropical fruits and seeds from South America and the Caribbean are regularly carried by the Gulf Stream to the rather inhospitable beaches of northern Europe. The most frequent arrivals from the New World are members of the legume family (Fabaceae), which is probably why people call them "sea beans". These seeds must have appeared strange to people throughout history, especially in the Middle Ages. It is not

Cephalophyllum loreum (Aizoaceae) – native to South Africa – fruit before and after wetting; the members of the stone plant family are rare specialists that rely on raindrops to flush the seeds out of their capsules. Rather than opening in dry conditions like most other capsules, their fruits open when they get wet and close again when they dry out. This hygrochastic opening mechanism ensures that the seeds are dispersed only when water is available for germination

opposite: *Cerbera manghas* (Apocynaceae) – pink-eyed cerbera; native from the Seychelles to the Pacific – drift fruit; commonly found as flotsam on beaches in the Indian and Pacific Oceans. Once the outer skin has rotted away, the cage of woody vascular bundles enclosing a massive corky mesocarp with large intercellular air-spaces becomes visible. The cork tissue ensures the fruit excellent, long-lasting buoyancy in sea water; fruit 9cm long

below: *Rhizophora mangle* (Rhizophoraceae) – red mangrove; photographed on the Pacific island of Pangai Motu in the Kingdom of Tonga – branch with three viviparous fruits (berries), one of which has already dropped its long green embryo; the plant in the background shows the typical growth form of the red mangrove

surprising that so many stories, legends and superstitious beliefs are woven around them. Christopher Columbus's voyage of discovery was allegedly inspired by the exotic sea-heart (*Entada gigas*) and the people of Porto Santo in the Azores still call the seed *fava de Colom* (Columbus's bean). Today, collectors and creators of botanical jewellery value them for their beautiful shapes and colours. Apart from the sea-heart, the most famous sea beans are the true sea bean (*Mucuna sloanei* and *M. urens*), the sea purse (*Dioclea reflexa*), grey and yellow nickernuts (*Caesalpinia bonduc, C. major*), and Mary's bean (*Merremia discoidesperma*). Most are legumes, although Mary's bean belongs to the morning glory family (Convolvulaceae).

Entada gigas grows as an enormous liana in the tropical forests of Central and South America and Africa. Its seeds are one of the most commonly found drift diaspores on European beaches. With a diameter of up to 5cm, the heart-shaped brown seeds are themselves large but are also borne in the largest of all legume pods, up to 1.80m long. Sea-hearts and the large seeds of the related *Entada phaseoloides* from Africa and Australia were commonly carved into snuff boxes and lockets in Norway and other parts of Europe. In England, the seeds were used as teething rings and good luck charms to protect children at sea. Grey nickernuts were worn as amulets by the people of the Hebrides to ward off the Evil Eye. The seeds were said to turn black when the wearer was in danger. The most intriguing sea bean has to be the Mary's bean. Produced by a woody vine that grows in the forests of southern Mexico and Central America, the black or brown, globose to oblong seeds are 20-30mm in diameter and 15-20mm thick. Their hallmark is a cross formed by two grooves, hence their name "crucifixion bean" or "Mary's bean". To religious people this seed had a symbolic meaning. Having survived the ocean, it was believed to give protection to anyone who owned it. In the Hebrides, for example, a woman in labour holding a Mary's bean was assured of an easy delivery. The seeds were handed down as precious talismans from mother to daughter for generations.

Apart from its resilience to sea water, the most important pre-requisite of a diaspore to successfully travel the oceans is buoyancy. Most tropical diaspores do not float either in fresh or sea water. It is estimated that less than one percent of tropical seed plants produce fruits or seeds that drift in seawater for at least a month. Those specifically adapted to water dispersal possess various buoyancy devices. Seeds increase their specific gravity by not filling the entire cavity of the thick, woody seed coat (e.g. kemiri nut, *Aleurites moluccana*, Euphorbiaceae), by leaving an air-filled gap between their two cotyledons (e.g. *Entada* spp., *Mucuna* spp., *Merremia* spp., *Mora megistosperma*), or by producing lightweight cotyledonary tissue (e.g. legumes like *Dioclea* spp.). Seaworthy fruits may or may not combine these characteristics with air cavities in their pericarp, or a fibrous or corky coat. With just the latter, box fruits (*Barringtonia asiatica*, Lecythidaceae), common flotsam on the beaches of French Polynesia, remain buoyant for at least two years.

The coconut (*Cocos nucifera*, Arecaceae) combines a fibrous, spongy fruit coat with an air bubble in the endosperm cavity. The seed inside is protected by a thick, hard endocarp, making the fruit of the coconut a dry drupe rather than a nut. Although no one would call it a sea bean, the coconut is the classic example of an ocean traveller. Its excellent adaptation to sea dispersal has spread coconut palms throughout the tropics. The average maximum distance that a coconut can travel while still afloat and viable is 5,000 kilometres. When it finally becomes stranded on a beach it will germinate slowly once rainfall has washed off the salt collected during its journey. Since sea-sand retains hardly any moisture, the liquid endosperm inside the coconut provides a crucial water reserve from germination until the roots of the seedling reach fresh groundwater.

The most enigmatic of drift fruits is the Seychelles nut, the fruit with the largest seed in the world. Although not closely related, the Seychelles nut is similar to a coconut, hence its alternative names *double coconut* and *coco de mer*. Unlike the coconut, the Seychelles nut is not adapted to ocean dispersal. It cannot float when fresh and does not survive prolonged contact with sea water. In the fifteenth century, long before the Seychelles were discovered in 1743, the endocarps were washed up on the beaches of the Indian Ocean. Since most of them were found on the Maldives the species was given the somewhat misleading Latin name *Lodoicea maldivica*. The true distribution of this extraordinary palm tree is limited to two islands of the Seychelles, Praslin and Curieuse. The Seychelles nut is famous not only for its size but also for the rather suggestive shape of its fruits, which gave rise to various superstitions. Malay and Chinese sailors thought that the double coconut grew on a mysterious underwater tree similar to a coconut palm. In Europe, it was thought that the highly-prized fruits had medicinal properties and that their endosperm was an antidote to poison. Just how valuable the *coco de mer* was before the discovery of the Seychelles is described by Albert Smith Bickmore in his *Travels in the East Indian Archipelago*, published in 1869: "*To early sailors in the Indian Ocean, the Seychelles nut was only known from drift seeds washed ashore throughout the region. The prince of Ceylon, who is said to have given a whole vessel laden with spice for a single specimen of the double coconut, could have satisfied his heart's fullest desire if he had only known it was not rare on the Seychelles, north of Mauritius.*"

Self-dispersal

Rather than entrusting their seeds to wind, water or animals, some plants have developed mechanisms that enable them to disperse their seeds themselves, at least for a short distance.

Sea-dispersal or *anochory* is either by actively catapulting the seeds away (*ballistic dispersal*) or by burying the fruits in the ground (*geocarpy*).

Ballistic dispersal

Mechanisms used by plants to expel their seeds from their fruits can be caused either by passive (hygroscopic) movements of dead tissues or by active movements that are due to high pressure in living cells.

Passive explosives

Dry, dehiscent fruits such as capsules and follicles open gradually along pre-formed lines as the pericarp dies and dries out. The shrinking of its tissues eventually causes the fruit wall to rupture. This is usually a slow, gradual process but in some fruits the pericarp is specifically adapted to build up a high degree of tension, which is eventually released by a sudden explosion that expels the seeds. The underlying principle of this mechanism is based on the different orientation of elongated cells in neighbouring layers, which often consist of crossed, thick-walled fibres. As the cells dry out, they contract parallel to their longitudinal axis, causing the neighbouring layers to pull in different directions. The tension is finally released in an explosive torsion movement of the separating fruit fragments, which usually correspond to whole or half carpels. In some legumes it is the two halves of the single carpel which separate along their dorsal and ventral sides. By twisting in opposite directions at lightning speed, the carpel-halves fling out the seeds with great force. Common broom (*Cytisus scoparius*), gorse (*Ulex europaeus*), sweet peas (*Lathyrus odoratus*) and lupins (*Lupinus* spp.) are familiar examples displaying this behaviour. Usually the seeds are dispersed over a short distance. In gorse, they remain within the radius of the mother plant with only about 2 per cent travelling 2-2.5m. As is often the case, the distances are more impressive in the tropics. The fruit of *Tetraberlinia morelina*, an African legume tree at home in the rainforests of west Gabon and southwest Cameroon, aided by its great height, shoots its seeds up to 60m from the mother plant. This is the longest ballistic dispersal distance ever recorded.

A plant family in which the typical fruit is an explosively dehiscent capsule is the Euphorbiaceae. In temperate herbaceous members such as the petty spurge (*Euphorbia peplus*), sun spurge (*Euphorbia helioscopia*), dog's mercury (*Mercurialis perennis*) and annual mercury (*Mercurialis annua*) the fruit is composed of three carpels. When it explodes, the fruit disintegrates into six half-carpels, which churn out three seeds. The most remarkable example in this family is native to the tropics of the New World. Not only has the gynoecium of the sandbox tree (*Hura crepitans*) from South America and the Caribbean

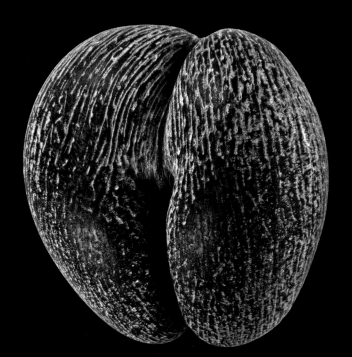

opposite: Sea beans – from top to bottom: *Merremia discoidesperma* (Convolvulaceae) – Mary's bean; collected in the New Hebrides, Scotland; c.2cm in diameter; *Caesalpinia bonduc* (Fabaceae) – nickernut; collected in Jamaica; 2.5cm long; *Entada gigas* (Fabaceae) – sea-heart; collected in Jamaica; up to c.4.5cm diameter; *Mucuna urens* (Fabaceae) – true sea bean or hamburger bean; c.2.5cm diameter

below: *Lodoicea maldivica* (Arecaceae) – Seychelles nut; native to the Seychelles – the single-seeded fruits, which take 7-10 years to mature, are produced by one of the most extraordinary palms and contain the largest seeds in the world. Before the discovery of the Seychelles, specimens were highly prized in Europe; fruit (fibrous mesocarp removed) 33cm long

above: *Euphorbia helioscopia* (Euphorbiaceae) – sun spurge; collected in the UK – seed with aril (elaiosome) to attract ants for dispersal. The arils of Euphorbias and their relatives are typically formed by a swelling of the outer integument around the micropyle; the central micropylar canal persists and is often visible in the mature seed; 2.3mm long

opposite: *Euphorbia epithymoides* (Euphorbiaceae) – cushion spurge; native to central and eastern Europe – close-up of inflorescence with developing fruits

many more (5-20) carpels than is usual in spurges. The ribbed fruit the size of a mandarin erupts much more violently, with the sound of a loud gunshot (Latin *crepare* = to burst). The force of the blast catapults the flat discoid seeds, of which there are as many as there are carpels, as far as 14m. The name sandbox tree dates back to the time before the invention of blotting paper and fountain pens, when the fruits served as containers for the sand used to blot the ink that ran profusely from goose quills.

Active explosives

Fleshy dehiscent fruits are able to build up sufficient pressure to explode by means of tissues of thin-walled cells, which increase their internal cell pressure (called *turgor*) by taking up additional water. As the tissues swell, at some point the pressure against neighbouring inelastic layers is so high that the slightest movement of the fruit sets off the explosion that expels the seeds. A passing animal can trigger the mechanism and although the seeds of such fruits are generally smooth and non-sticky, they may become entangled in its fur and be carried some distance. Classic examples of actively exploding fruits are the touch-me-not (*Impatiens* spp.) in the balsam family (Balsaminaceae), and the squirting cucumber (*Ecballium elaterium*) in the gourd family (Cucurbitaceae).

Anyone who has played with the fruits of the touch-me-not, gently squeezing them between two fingers, knows that its name is appropriate. Its pendulous fruits are composed of five carpels, with pre-formed dehiscence lines between them. Pressure builds up in the outer layers of the pericarp and when it is strong enough to separate the carpels, the slightest movement causes the fruits to explode. The outside of the pericarp expands more than the inside causing the fruit valves to curve inwards in a lightning-fast movement that hurls the seeds up to 5m away. The trigger for the explosion can be a passing animal, raindrops or water dripping off a tree, the wind, or even seeds launched from a neighbouring fruit. As the fruits of the Mediterranean squirting cucumber ripen, the inner tissue layers build up turgor pressure against the thick, rather inelastic outer skin. The pre-determined breaking line at the base of the stalk points up in the air as the fruit turns almost 180° against the pedicel. Eventually, the slightest movement causes the stalk to pop out like a cork, leaving a hole through which both watery juice and seeds squirt out. The pressure inside the fruit is so high that the seeds can travel more than 10m.

The true record breakers are dwarf mistletoes, which apply more or less the same principle as the squirting cucumber. Predominantly native to North America, they are parasites on pine trees and belong to the genus *Arceuthobium* in the Christmas mistletoe family (Viscaceae; recently united with the sandalwood family, Santalaceae). Whereas other

above: *Lathyrus clymenum* (Fabaceae) – jointed pea; native to the Mediterranean – close-up of the dehisced fruits; explosively dehiscent legume fruits eject their seeds by twisting the two halves of their single carpel in opposite directions

below: *Hura crepitans* (Euphorbiaceae) – sandbox tree; native to South America and the Caribbean – the ribbed fruit of the sandbox tree is the size of a mandarin and erupts violently to catapult its flat discoid seeds as far as 14m

opposite: *Impatiens glandulifera* (Balsaminaceae) – Himalayan Balsam; native to the Himalayas – close-up of flowers and fruits (above) and a fruit that has exploded (below). When the seeds are ripe, pressure builds up in the outer layers of the fruit wall (pericarp). Eventually the slightest movement causes the fruits to explode and the seeds are hurled up to 5m away

mistletoes rely upon birds for the dispersal of their seeds, dwarf mistletoes (with the exception of *Arceuthobium verticilliflorum*) have evolved explosively dehiscent fruits. Just as in the squirting cucumber, the pedicel bends downwards as the fruit matures and a dehiscence line forms around its point of attachment. Once the fruit has reached maturity, the slightest touch dislodges it from the pedicel. The pressure that builds up inside the dark green, single-seeded berry is so high that it fires the tiny sticky seed through the hole in the rubbery fruit wall over a distance of up to 16m at the remarkable speed of 2m per second (97km per hour). A single ponderosa pine infected with *Arceuthobium campylopodum* may be bombarded by more than two million seeds, which overwinter on the surface where they land and germinate the following spring. If a seed is lucky enough to land on a compatible host tree, its embryo grows, supported by the photosynthetic endosperm tissue, and enters the host's bark. *Aceuthobium* species can live inside their host for months or even years before producing a plant.

Whether the seeds are catapulted from passively or actively exploding fruits, they are dispersed over no more than a few metres. The advantage of ballistic dispersal is that it is cheap, requiring no animal reward and usually very little in terms of specialised structures. The preferred strategy of annual plants is to disperse their seeds in a way that allows them to maintain and expand an existing population in the same place rather than colonising new territories. For perennial plants the most important factor is that the seeds should avoid competition with the parent plant. However, the ejection of the seeds from their fruits is often only the first phase of their dispersal history. Many seeds, such as the squirting cucumber, pansies, spurges and gorse, have an energy-rich, oily appendage, which functions as an edible bait that lures ants into carrying them away from the parent plant. Apart from such appendages, ballistically dispersed seeds are usually small, smooth and more or less spherical, ensuring low air resistance.

Geocarpy – or how do peanuts end up underground?

This is a legitimate question if one has never seen a peanut plant (*Arachis hypogaea*, Fabaceae). After all, peanuts are fruits and fruits develop from flowers, and flowers need to be pollinated, which mostly happens in the air. So how can a fruit end up underground? The answer is that the plant itself buries them. After pollination and fertilization, the ovary is pushed down into the ground, hence the name *geocarpy* for this type of self-dispersal. The single carpel of the peanut flower sits at the end of a specialised stalk-like organ called the *gynophore*. As soon as the ovules are fertilised, the gynophore bends down and elongates until it has pushed the young, pointed ovary underground. Once the ovary has reached its

subterranean destination, it swells to produce peanuts. Other legumes that bury their fruits in the same way are the Bambara groundnut from West Africa (*Vigna subterraneana*) and *Astragalus hypogaeus* from west Siberia.

Geocarpy is mostly found in annual plants living in dry, hot climates such as deserts, grasslands and savannah habitats. Here, burying the fruits is not only a safeguard against grazing animals, it primarily ensures that the next generation is kept in a suitable location in an inhospitable environment.

One rare temperate example of a plant that buries its fruits is the ivy-leafed toadflax (*Cymbalaria muralis*, Plantaginaceae), a common wild flower on walls and rocks in Britain, Europe and North America. When in bloom, the flowers face the sun. As soon as they have been pollinated, their pedicels turn away from the sunlight and seek out shade by growing longer and searching the substrate for suitably dark cracks and crannies in which to deposit their fruit. When ripe, the capsules open to release a number of small, irregularly ornamented seeds. Careful planting by the mother plant and the rugged surface of the fruits prevent them from rolling out of their protected environment.

Self-burying drills

Some diaspores bury themselves after dispersal. Apart from keeping them hidden from animal predators, this self-burial is regarded as an adaptation to dry soils as it allows the diaspore to reach the more humid layers just under the soil surface. The rotating movement of their hygroscopic appendages, which twist and untwist with changes in humidity, allows the diaspore to drill itself into the ground. This behaviour is most famously found in species of stork's bills, e.g. red-stem stork's bill (*Erodium cicutarium*), and musky stork's bill (*E. moschatum*). Their diaspores consist of fragments (fruitlets) of a schizocarpic fruit with a long beak. Once the fruit has come apart, each single-seeded fruitlet retains a share of the beak, which serves as a hygroscopically moving awn. The same mechanism is found in a number of species of the very distantly related grass family (Poaceae). The florets of wild oat (*Avena fatua*), wild barley (*Hordeum vulgare* ssp. *spontaneum*) and the needle-and-thread grass (*Stipa comata*) are equipped with an awn, the basal portion of which twists and untwists with changes in moisture levels. The drilling florets of the Australian corkscrew spear-grass (*Stipa setacea*) are so sharp that they are able to penetrate the wool and skin of sheep. Once embedded in the flesh of the animal, the curved hairs on the surface of the florets make them almost impossible to dislodge. The muscle movements of the tortured animals drag the fruits deeper and deeper into their body. Florets of *Stipa setacea* have allegedly been found in the heart muscle of dead sheep.

Cymbalaria muralis (Plantaginaceae) – ivy-leaved toadflax; collected in the UK – seed with no obvious adaptations to a specific mode of dispersal. The seeds are planted by the mother plant, which deposits its fruits in dark cracks and crannies. The rugged surface of the seeds may prevent them from rolling out of their sheltered location

Creeps and jerks

Hygroscopically moving appendages enable other diaspores to creep along the ground for short distances. Such creeping diaspores are found in some grasses, and in the sunflower family (Asteraceae) and teasel family (Dipsacaceae). Their fruits have a hygroscopically moving pappus (modified calyx). For example, the fruits (cypselas) of the cornflower (*Centaurea cyanus*, Asteraceae) are crowned by a tuft of short, stiff, scale-like pappus segments, which are too small to play any role in wind dispersal. Their function is rather different: with changing humidity the pappus scales repeatedly move in and out, thus pushing the fruits a few centimetres over the ground. The very short, forward-pointing teeth along their margins prevent movement in the opposite direction. The distances they creep are short but at least they move away from the parent plant; wind and rainwater may carry them further. A more targeted additional strategy has given them an edible swelling at the base to specifically attract ants, a mode of dispersal that will be discussed in detail.

Specialised awns enable the fruits of certain grasses (e.g. *Arrhenatherum elatius*, *Avena sterilis*) to jerk and jump for short distances. The distal part of their long-kneed awns is straight whereas the lower part is helically twisted and extremely hygroscopic. With changing humidity the basal part winds or unwinds, turning the straight distal part of the awn. Since each fruit has two awns, which turn in opposite directions, their distal parts eventually meet and become entangled. The tension that builds up between them is released when the pressure is strong enough to push the distal parts past each other. Within a split second, the jerky movement of the awns catapults the diaspore into the air.

Dispersal by animals

Animal movements are less haphazard than wind and water making animals much more reliable as dispersal agents. A plant that manages to develop a relationship with animals needs fewer seeds to guarantee the survival of the species. The evolution of animal-dispersed diaspores clearly has many advantages and so it is not surprising that some fifty per cent of gymnosperms (*Ephedra*, *Gnetum*, *Ginkgo*, a few conifers and all cycads) use animals to assist in the dispersal of their seeds. Once again it was the angiosperms that perfected the shift from abiotic to biotic dispersal agents by evolving a fascinating spectrum of strategies enabling them to travel either on or inside animals. These close, sometimes highly specialised relationships between angiosperms and animal dispersers provide another key to the understanding of the evolutionary success of this group.

The evidence of such strategies in plants lies in every sweet, juicy fruit we enjoy. The sweet pulp of the fruit is a bait to lure potential dispersers into swallowing the seeds and

Centaurea cyanus (Asteraceae) – cornflower; collected in the UK – fruit (cypsela); the tuft of short, stiff, scale-like pappus segments at the top is too small to play any role in wind dispersal. Instead, with changing humidity the scales repeatedly move in and out, pushing the fruits over the ground for a few centimetres as they do so. Movement in the opposite direction is avoided by very short, forward-pointing teeth arranged along the margin of the pappus scales. To ensure further dispersal by ants the cypsela has an edible elaiosome at the base

The Dispersal of Fruits and Seeds

dispersing them in their faeces. If they are fortunate, the seeds will land in a suitable place to grow well away from the shadow of the mother plant. This form of dispersal is called *endozoochory*, "dispersal inside an animal". Endozoochory is a frequently encountered phenomenon in many families with fleshy fruits, e.g. Ericaceae, Rosaceae, Solanaceae. Endozoochorously dispersed seeds do not present any conspicuous adaptations to facilitate their dispersal. They are usually smooth, globose to ovoid, and either covered in a hard endocarp (if borne in drupes) or with a hard seed coat (if borne in berries) to withstand gastric juices and intestinal enzymes. Many seeds from fleshy fruits germinate better after passing through the gut of an animal.

Among vertebrates, the most important dispersers are birds and mammals, especially in temperate climates. In the tropics, where dispersal is accomplished mostly by vertebrates, fish and reptiles also act as dispersers. Vertebrate-dispersed diaspores offer a reward in the form of a nutritious pulp rich in sugars, protein or fat, but only when the seeds are mature. Ripening fruits are inconspicuous in colour, rather hard, and have no smell; at best they are sour, at worst poisonous, and always unpalatable. As soon as the seeds are ripe, the fruit sends out signals of the promise of a safe, nutritious reward. The nature of the signals depends on the kind of animal that is to be attracted. Birds have excellent colour vision but only a poorly developed sense of smell. Diaspores adapted to bird-dispersal (*ornithochory*) are therefore odourless but may change colour from green to something more conspicuous such as red, which is the colour that birds distinguish best against a green background; but purple, black and sometimes blue or combinations of these (especially red and black) are also found. For mammals, a slightly different strategy promises greater success. Mammals rely much more on their keen sense of smell and many are nocturnal. Mammal-dispersed fruits are therefore often (but not always) dull in colour (brown or green) and exude a strong aromatic scent when ripe. Apples, pears, medlars, quinces, rosehips, citrus fruits, mangoes, papayas, passionfruits, melons, bananas, pineapples, jackfruits, breadfruits and figs all target mammals such as rodents, bats, bears, apes, monkeys and even elephants as dispersers of their seeds. Sometimes the plants rely entirely on animals. The fruits of *Gardenia thunbergia* (Rubiaceae) from eastern South Africa, which are greyish-green, hard and inconspicuous, can remain on the shrub for several years if they are not eaten by antelopes (or used as tool handles by indigenous people).

Compared with birds, mammals play a minor role as seed dispersers in the temperate climate of Europe where berry-eating animals include not only plant-eaters such as deer and wild boars but also some carnivores such as foxes, martens and badgers. In the tropics mammal-dispersed fruits are much more common. Apes, monkeys, and particularly fruit bats

below: *Paeonia cambessedesii* (Paeoniaceae) – peonia; native to the Balearic Islands – multiple fruit; when ripe, the individual follicles open to display numerous fertile, shiny black, pea-size seeds and a number of smaller, bright red, fleshy sterile seeds. The purpose of the latter is an improved visual signal and edible reward for visiting birds

opposite top: *Podocarpus lawrencei* (Podocarpaceae) – mountain plum pine; native to southern Australia and Tasmania – fruit, c.10-14mm long; the name *Podocarpus* is derived from the Greek (*podos* = foot + *karpos* = fruit) and refers to a characteristic of plum pines: the single seed is subtended by a brightly coloured fleshy swelling (aril) that attracts animals and especially birds for dispersal

opposite middle: *Taxus baccata* 'Fructo luteo' (Taxaceae) – a horticultural variety of the English yew with yellow, rather than red fruits. Yew "berries" consist of a single seed surrounded by an 8-15mm long cup-shaped aril. The aril is mature 6-9 months after pollination and is the only part of the plant that is not poisonous. Yew fruits are eaten by birds (e.g. thrushes, waxwings), which disperse the seed undamaged with their droppings

opposite bottom: *Gardenia thunbergia* (Rubiaceae) – wild gardenia; native to South Africa – the egg-shaped, hard, fibrous fruits are about 5-7.5cm long and 3.5cm in diameter. Their natural dispersers are elephants, antelopes and buffalos (hence their Afrikaans name "buffelsbal"). If the fruits are not eaten by animals they can stay on the plant for years

are among their primary dispersers. Nevertheless, the most important endozoochoric dispersers of all are birds. This may be due in part to the large number of bird species. They are also popular dispersers because they do not chew their food, which means that they do not destroy the seeds as they devour the fruit. Ornithochorous fruits are generally not very large and contain either one large stone or seed that is not swallowed by the bird (e.g. cherries) or many small seeds, which are ingested with the fruit pulp (e.g. currants). In a temperate climate, bird-dispersed fruits mature in the autumn, when birds need to stock up on their energy reserves before the long migration south.

Seeds seeking attention

Although animal-dispersed diaspores are usually soft fruits, such as berries and drupes offering nutritious fruit pulp, in some plants the reward is the seed itself. The edible part can be either a fleshy seed coat, called a *sarcotesta*, or a special appendage, called an *aril*.

Fleshy seed coats

When the seed coat provides the reward, only the outer parts become fleshy and produce a conspicuously coloured sarcotesta; the inner layers usually form a hard protective shell, called a *sclerotesta*. Sarcotestal seeds were already present among the gymnosperms, most notably in *Gnetum* and the archaic *Ginkgo* and cycads. The presence of sarcotestal seeds in these living fossils suggests that offering an edible seed coat to attract animal dispersers is an ancient strategy that probably led seed plants to their long, successful interaction with animals as distributors of their diaspores.

With their thick, orange-yellow sarcotesta *Ginkgo* seeds resemble a yellowish berry or drupe. Despite its rather appetizing appearance, the fleshy part of the ripe seed contains butyric acid and exudes the infamously foul smell of rancid butter as it rots after the seeds have dropped. It is not known which animal was to be enticed by this smell. Since ginkgos appeared long before mammals their most likely natural dispersers were probably dinosaurs. Their bad smell makes female *Ginkgo* trees unwelcome in ornamental planting. Nevertheless, in Asia the (cleaned) seeds, called ginkgo nuts, are considered a delicacy. Their edible part, the massive megagametophyte tissue inside the sclerotesta, is eaten boiled or roasted in China, Japan and Korea; it is also used in traditional Chinese herbal medicine to treat lung problems, and in the West to improve the memory. Cycads reveal their large, often brightly coloured seeds as the megasporophylls separate and the cones disintegrate at maturity. The sarcotesta is mostly red, purple or scarlet (e.g. *Encelphalartos longifolius, Lepidozamia hopei, Macrozamia fraseri, Stangeria eriopus*) but can be orange (e.g. *Encephalartos*

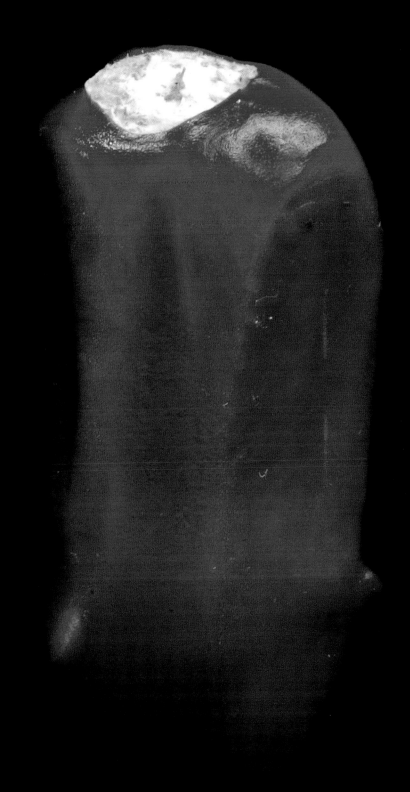

laevifolius, Zamia furfuracea), brown (e.g. *Encelphalartos dyerianus*), yellow (e.g. *Cycas macrocarpa*) or white (*Ceratozamia* spp., *Dioon* spp.). In many species of *Encelphalartos* the mature female cones are themselves highly coloured (e.g. *E. ferox* with reddish cones), undoubtedly an adaptation to attract potential animal dispersers. The fleshy sarcotesta of cycad seeds (and sometimes the whole seed) provides food for a wide variety of animals such as birds (e.g. parrots, cockatoos, hornbills, emus, cassowaries), rodents (e.g. rats, mice, squirrels), small marsupials (e.g. possums, kangaroos, wallabies), fruit bats, and some larger mammals (e.g. bears, peccaries, baboons, monkeys).

The sarcotestal seeds of angiosperms are primarily bird-dispersed. As the green follicles of the multiple fruits of magnolias open, each releases one or two bright red seeds which are suspended from their follicles on thin silky threads and dangle in the slightest breeze to attract birds. The origin of these silky threads is most unusual: as the seed drops out of its follicle the helical wall-thickenings present in the vascular elements of the funiculus unroll and abseil the seed. Similar fruits and seeds are found in peonies (*Paeonia* spp., Paeoniaceae). In *Paeonia cambessedesii*, the individual follicles of the multiple fruits open when ripe to display numerous pea-size seeds with a shiny black sarcotesta. To enhance both the optical contrast and the edible reward for the birds, smaller sterile seeds, which are fleshy and bright red in colour, are interspersed between the fertile black seeds. A popular fruit whose pulp consists of the fleshy outer cell layer of the seed coat is the pomegranate (*Punica granatum*, Lythraceae). Originally native to an area from Iran to the Himalayas, the pomegranate has long been cultivated and naturalized in the Mediterranean region for its fruits, especially in the Middle East. The seeds can be eaten straight from the fruit or used to prepare a juice or syrup (grenadine). The Tree of Knowledge in the Garden of Eden was probably a pomegranate. Nowhere is it stated that it was an apple.

Edible seed appendages

Seeds also use edible appendages (*arils*) to attract birds and other animals. Among the gymnosperms, arillate seeds are found only in certain conifers, which bear their seeds in single-seeded, berry- or drupe-like structures. Best known among these are the members of the yew family (Taxaceae) and yellowwood family (Podocarpaceae) in which the mature seeds are either subtended by a bright red fleshy aril or partly or entirely covered by it. The common name "yew berry" refers to the fleshy red cup surrounding the seed of the English yew (*Taxus baccata*); the sweet-tasting aril is the only part of the plant that is not poisonous. The fruits of *Ephedra* in the Gnetales are superficially similar but their fleshy outer part is formed by two pairs of bracts (modified leaves).

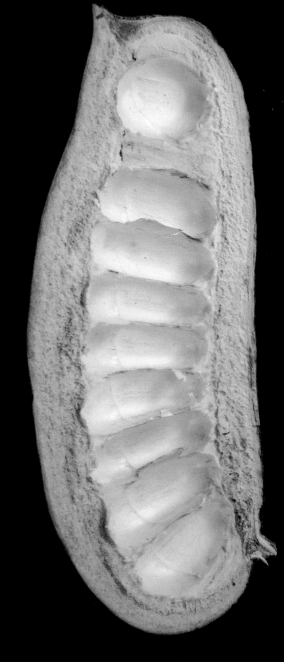

previous pages: *Encephalartos longifolius* (Zamiaceae) – native to South Africa – as in all cycads, the female cones disintegrate at maturity shedding their megasporophylls, which each bear two seeds. The seed cones of this species are among the largest of any *Encephalartos* species and weigh as much as 40kg. The Hottentots used to bury the stems for up to two months and then use their pith to prepare a form of bread; seed c.4-5cm long

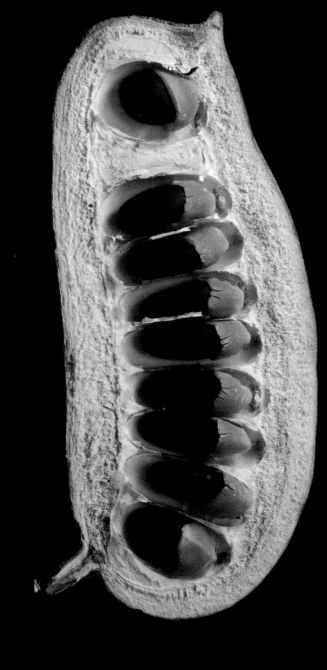

opposite and above: *Afzelia africana* (Fabaceae) – African mahogany; collected in Burkina Faso – open fruit (legume) consisting of the two halves of the single carpel; inside the fruit are a number of large black seeds with bright orange-red arils to attract birds for dispersal. Because of their attractive appearance the seeds are also used by makers of botanical jewellery; pod 17.5cm long

An infinitely larger variety of seed appendages is found in the angiosperms. Their arils evolved convergently and are of a totally different origin from those of the conifers. Arillate seeds in the angiosperms are adapted to attract both vertebrate and invertebrate animals. They are common in the legumes (Fabaceae) where they are formed by a swelling of the funiculus at the hilum of the seed. The large black seeds of the African mahogany (*Afzelia africana*), for example, have a highly nutritious aril at the hilar end, its bright orange-red colour proving a showy attraction. After nibbling off the aril, birds discard the large seeds. Because of their attractive appearance, the seeds are thought of as lucky charms and used in botanical jewellery. An interesting variation of a funicular aril is found in some *Acacia* species. The black seeds of *Acacia cyclops* from Western Australia are surrounded along the periphery by a vastly elongated funiculus, which turns into a bright orange-red fleshy aril when the seeds are ripe. Six times richer in fat than those of other acacias, the arils of *Acacia cyclops* are an important energy source for birds. In South Africa, where the species has become an invasive weed, the nutritious funicles are also eaten by rats and baboons. A native species of South Africa with rather spectacular arillate seeds is the bird-of-paradise flower (*Strelitzia reginae*, Strelitziaceae). As is typical of many Monocots, its three-carpellate gynoecium develops into a loculicidal capsule, which opens with three valves. Lined up in two rows along each valve are the pea-size black seeds, which attract the attention of birds by means of an appendage that looks like a shaggy bright orange wig with thick oily hairs that originate partly from the funiculus and partly from the outer integument around the micropyle. It is very difficult to find an open fruit with seeds still inside – proof of the great success of this strategy.

Some of our most cherished tropical and subtropical fruits are so delicious because their juicy seed appendages are designed to attract mammals rather than birds. Many originate in South-East Asia, such as the lychee (*Litchi chinensis*) and the longan (*Dimocarpus longan*), both members of the soapberry family (Sapindaceae). Their sweet flesh consists of funicular tissue wrapped round a single large shiny brown seed. The ovary wall contributes only the thin leathery outer skin. Sometimes called *Queen of Fruits* or *Food of the Gods*, the fruit of *Garcinia mangostana* (Clusiaceae) is better known as mangosteen. The indehiscent berry-like fruits are the size of a mandarin or small orange. Hidden inside the thick, dark purple, leathery and unpalatable rind sit 5-10 flat seeds wrapped in a thick juicy, snow-white to pinkish funicular aril. The flesh of the aril has a delicious smell and an exquisite taste – described variously as a blend of pineapple and peaches or reminiscent of strawberries mixed with oranges. Unfortunately, the fruits last only a couple of days and are best enjoyed straight from the tree. Those imported by supermarkets overseas have to be harvested in an unripe state to the detriment of their taste. Some dexterity is required to get through the thick leathery rind of the mangosteen and other mammal-dispersed fruits such as oranges and bananas to access the delicious contents. Best equipped for this are monkeys, who are also the natural dispersers of the mangosteen in its native South-East Asia. The seeds are rarely swallowed but rather discarded once the fleshy aril has been nibbled off.

If the mangosteen is the *Queen of Fruits*, the durian (*Durio zibethinus*, Malvaceae) has to be the *King of Fruits*, at least for the people of South-East Asia. One durian can weigh almost three kilograms and be as large as a football. Ferociously spiny on the outside, the greenish capsule bears a number of seeds wrapped in a large, delicious funicular aril. When the fruits are ripe, the hard tissue of the yellowish aril becomes a custard-like cream the consistency and flavour of which have been described as a tantalising mix of nuts, spices, bananas, vanilla and onions. After his first visit to Borneo, the nineteenth-century naturalist Alfred Russell Wallace wrote: "*Its consistence and flavour are indescribable. A rich butter-like custard highly flavoured with almonds gives the best general idea of it, but intermingled with it come wafts of flavour that call to mind cream-cheese, onion sauce, brown sherry, and other incongruities.*" Among the natural dispersers of the durian are orang utans, Asian rhinos, tapirs, bears and elephants. What attracts these animals is not the dull colour of the fruit but its pervasive smell. The durian is the most pungent fruit in the world and smells of a devious blend of faeces and rotten garlic. Most people in South-East Asia delight in the taste of the creamy arils but Westerners are usually put off by the stench. The strategy of large arils is to attract vertebrate dispersers but there is an entire microcosm of small seeds and other diaspores that bear fatty nodules designed to recruit much smaller couriers – ants.

Dispersal by ants

In 1906 the Swiss biologist Rutger Sernander first pointed out that the seeds of many plant species are specially adapted to dispersal by ants and called this strategy *myrmecochory*. Myrmecochorous seeds are small and light enough for ants to carry them back to their nests. As a reward for their efforts, they receive food in the form of an appendage that is rich in fatty oil and contains sugars, proteins, and vitamins (the *elaiosome*). The ants use these highly nutritious nodules, called elaiosomes mostly to feed their own brood. They keep the seeds in the nest until the nutritive tissue has been consumed and then usually either abandon the still viable seeds inside the nest or discard them outside on the rich soil of the colony's waste pile. Much of this seed-carrying behaviour is stereotypical and is induced by the presence in the elaiosome of ricinolic acid, the same unsaturated fatty acid that is found in the secretions of ant larvae.

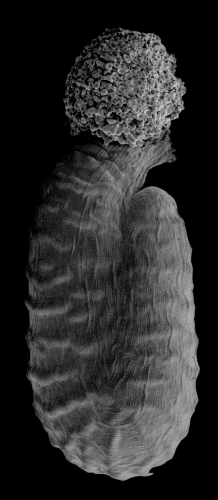

above: *Tersonia cyathiflora* (Gyrostemonaceae) – button creep collected in Western Australia – seed with elaiosome to assist dispersal by ants; seed (including elaiosome) 2.7mm long

top: *Codonocarpus cotinifolius* (Gyrostemonaceae) – desert po collected in Western Australia – seed with elaiosome to assist dispersal by ants; seed (including elaiosome) 2.6mm long

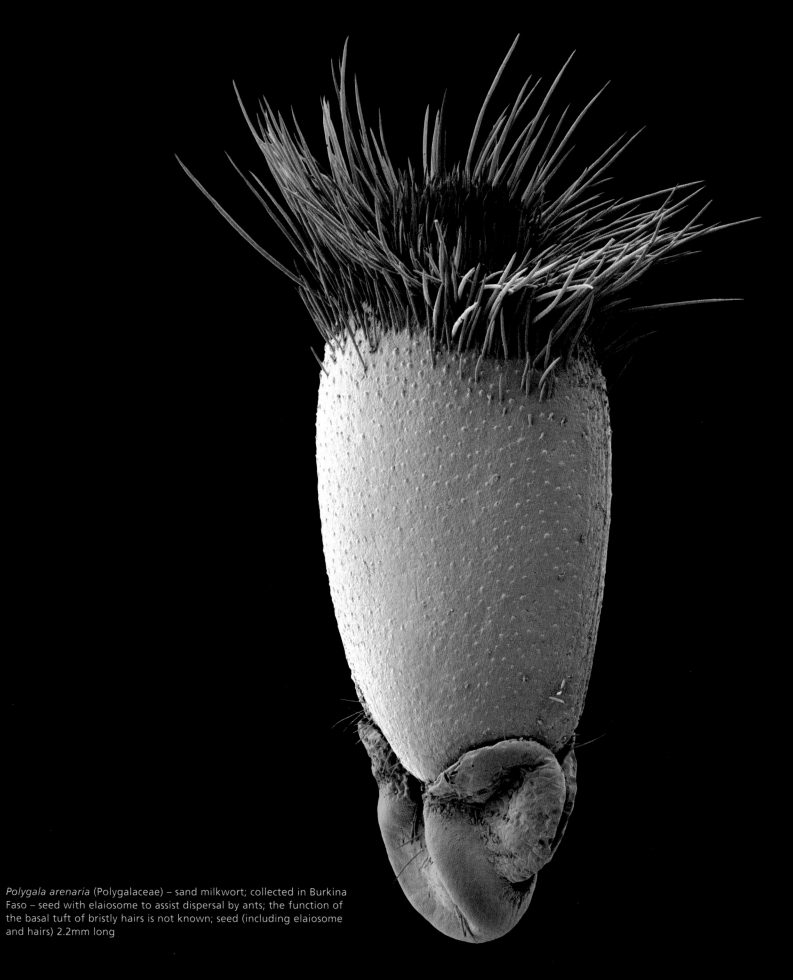

Polygala arenaria (Polygalaceae) – sand milkwort; collected in Burkina Faso – seed with elaiosome to assist dispersal by ants; the function of the basal tuft of bristly hairs is not known; seed (including elaiosome and hairs) 2.2mm long

The Dispersal of Fruits and Seeds

Morphologically, elaiosomes can be of various origins. Often they are seed appendages (arils) derived from different parts of the seed or funiculus. The elaiosomes of the seeds of gorse (*Ulex europaea*, Fabaceae), some cacti (e.g. *Aztekium, Blossfeldia, Strombocactus*) and many Caryophyllaceae (e.g. *Moehringia trinerva*) are swellings of the funiculus at the hilum. The seeds of *Glinus lotoides* (Molluginaceae) have a strange funicular aril consisting of two lateral lobes with a tail-like extension in between. In the seeds of many members of the spurge family (*Euphorbia* spp., *Mercurialis* spp., *Ricinus communis*; Euphorbiaceae) and milkweeds (e.g. *Polygala* spp., Polygalaceae) the elaiosomes are formed by an outgrowth of the outer integument around the micropyle (*exostome* arils). The fatty appendages of the seeds of the greater celandine (*Chelidonium majus*, Papaveraceae) are produced by a swelling of the raphe; and in pansies both the raphe and the exostome participate in the creation of the elaiosome. In other cases, elaiosomes are produced by the fruit wall, as in the cypselas of some Asteraceae (e.g. *Centaurea cyanus*) or the nutlets of some Lamiaceae (e.g. *Lamium maculatum*) and Boraginaceae (e.g. *Borago officinalis, Pulmonaria officinalis, Symphytum officinale*). These are just a few examples of some of the ways in which elaiosomes are formed in ant-dispersed diaspores.

Myrmecochory is found in over eighty plant families. It plays a major role in temperate deciduous forests in Europe and North America and especially in the dry shrublands of Australia and South Africa. Australian heathlands are home to more than 1,500 ant-dispersed plant species. The average dispersal distance of myrmecochorous seeds is very short – between one and two metres from the parent plant, although it can be 70m in extreme cases. The benefits of this animal-plant relationship are mutual: the ant is given a nutritious, reliable source of food while the plant's seeds are carried far enough to reduce competition between seedlings and parent. By burying the seeds a short distance below the soil surface, the ants hide them from predators such as rodents, and also prevent them from being destroyed by fire. The latter may explain the importance of ants as dispersal agents in the dry habitats of Australia and Africa that are regularly swept by seasonal field fires.

Ant dispersal seems to be obligatory for some plants whereas in many others it acts as a supplementary distribution mechanism following ballistic dispersal. The former applies to cyclamen. Once the flowers have been pollinated, the stalks bend down and curl up in a tight spring pulling the developing seed capsule to the ground. Strategically placed at ground level, the ripe capsules open with five to seven valves presenting their sticky seeds, ready to be collected by ants attracted by their sugary outer layer. Often myrmecochory is combined with autochory (self-dispersal). The capsules of the greater celandine (*Chelidonium majus*, Papaveraceae) open with two pores at the base through which the

Harvester ant (*Messor* sp., Formicidae) carrying a fruit to its nest; photographed in Greece

elaiosome-bearing seeds drop to the ground to be collected by ants. In many other species, myrmecochory is combined with explosively dehiscent fruits as is the case in many legumes (e.g. gorse, *Ulex europaea*), pansies (Viola ssp.), spurges (*Euphorbia* spp.), and mercuries (*Mercurialis* spp.). How reliant certain myrmecochorous plants become on their specific ant dispersers is revealed by an example from the Cape in South Africa. Many plants of the Cape fynbos vegetation have myrmecochorous diaspores. Two decades ago, the dominant native ants in the Cape region were displaced by the invasive Argentine ant, *Iridomyrmex humilis*. Unlike the larger native ants, the much smaller Argentine ants move the diaspores only for very short distances and do not store them underground in their nests. Consequently, the diaspores (achenes) of the Proteaceae, *Mimetes cucullatus* are left exposed and eaten by predators. The result is that the seedlings are reduced in number and grow ever closer to the parent plant. In the long term this invasion may lead to the extinction of many members of the spectacular Cape flora. Recent research reveals that this is not a unique scenario: small invasive ants such as the Argentine ants and fire ants are moving into ecosystems throughout the world – for example in the United States and Australia – where they are displacing the traditional local ant partners of many native plants.

Dispersal by scatter-hoarders

A similar strategy to ant dispersal but one that requires no obvious adaptation of the diaspores is to rely on the bad memory or premature death of scatter-hoarding predators. Rodents and certain birds store seeds as food for the winter. Some seeds have to be sacrificed to provide a meal for the animal, but squirrels, for example, never find all the seeds they have buried, thus leaving a good number to germinate. This strategy has proved so successful for certain plants that their fruits have been adapted to suit a single scatter-hoarding animal. The reliance of the Brazil nut tree on agoutis has already been described. A European example is the similarly exclusive relationship between the gymnospermous arolla pine (*Pinus cembra*, Pinaceae) and the appropriately named nutcracker bird, which plays a vital role in the life cycle of the arolla pine at high altitudes in the central Alps. With its very strong, specialised beak the nutcracker is almost alone in being able to dig the seeds out of the tough cones and crack their hard shells. It collects them avidly for the winter, storing up to sixty in its crop at any one time. Once it has a good harvest, it regurgitates the seeds and hides them underground, far from the original tree. The nutcracker has a phenomenal memory and will find its buried seeds even when they are covered by a metre of snow but at least a few of the 100,000 seeds it collects in a good year will escape its scrutiny and give rise to new trees.

Waste pile outside a nest of harvester ants (*Messor* sp., Formicidae) in Greece; among the debris are fruit remains of *Medicago disciformis*, *Trifolium campestre* (hop trefoil), *Medicago polymorpha* (burr medic), *Scorpiurus muricatus* (caterpillar plant), *Astragalus hamosus* (all Fabaceae) and *Silene vulgaris* (bladder campion, Caryophyllaceae)

Sticky hitchhikers

Rather than offering expensive rewards or sacrificing part of their seeds to bribe animals to disperse their fruits and seeds, many plants prefer hitchhiking as a cost effective alternative means of travel. They have developed spines, hooks or sticky substances, which they use to attach themselves to the skin, feathers, fur or legs of unsuspecting mammals and birds. Since the diaspores ride on the outside of the animal, this dispersal mode is called *epizoochory*. A great advantage of epizoochory is that the dispersal distance is not limited by factors such as gut retention times, as in endozoochory. Eventually, most adhesive diaspores fall off by themselves or are removed by the animals to which they are attached. Some lucky passengers remain undetected or settle in a spot where they cannot be groomed away and stay for a long time with the animal, sometimes until it dies. Epizoochorous diaspores therefore stand a good chance of being dispersed over greater distances than most diaspores relying on other modes of dispersal.

Seeds that use a mucilaginous coat to stick to a passing animal are rare. The few examples include linseed (*Linum usitatissimum*, Linaceae), plantains (*Plantago* spp., Plantaginacae), members of the mustard family (Brassicaceae, e.g. *Lepidium* spp.), and parasitic mistletoes. The most common and very effective adaptation of epizoochorous dispersal is to cover the surface of the diaspore with barbed spines or hooks. After a walk through the woods we have all experienced the tenacity of burrs. Among the stickiest examples in temperate climates are the fruits and even the entire capitulae (seed heads) of members of the sunflower family (Asteraceae), the most famous being beggar's ticks (*Bidens pilosa*), the burdock (*Arctium lappa*) and the cocklebur (*Xanthium strumarium*). Other common burrs are the fruitlets of the schizocarpic fruits of goosegrass (also called sticky-willy; *Galium aparine*, Rubiaceae), houndstongue (*Cynoglossum officinale*, Boraginaceae) and American stickseed (*Hackelia deflexa*, Boraginaceae).

The first to grasp the remarkable mechanism of these sticky diaspores was the Swiss electrical engineer George de Mestral, an amateur mountaineer and naturalist. One day in 1948 after a walk in the Jura mountains he took a close look at the burrs sticking to his hunting trousers and his dog's fur. He found that the adhesive principle was based on small hooks that clung to the tiny loops of thread in the fabric of his trousers. Inspired by nature's hook and loop principle, the engineer invented a new fastener. Despite public resistance and ridicule, de Mestral continued to perfect his design for a new fastener and finally, in 1955, patented Velcro® (based on the French *velour* = velvet and *crochet* = hook) and started his own company. When he sold Velcro Industries it was producing fifty-five million metres a year of the famous hook-and-loop fastener.

Hackelia sp. (Boraginaceae) – stickseed; collected in Canada – single-seeded nutlet displaying the most common and effective strategy to ensure dispersal on an animal, namely to cover the surface of the diaspore with barbed spines or hooks. As in many members of the borage family, the ovary of *Hackelia* is deeply four-lobed and splits at maturity into four single-seeded nutlets; each nutlet (including spines) 3.5mm long

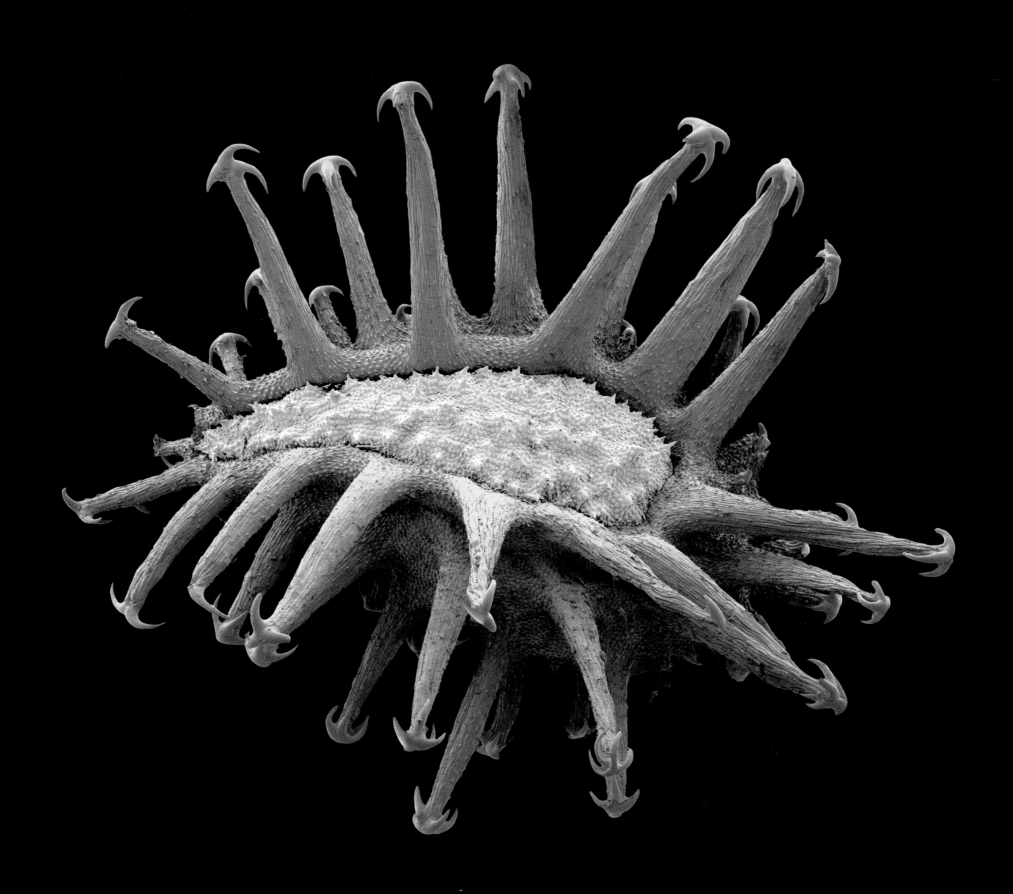

The Dispersal of Fruits and Seeds

comparison with the ruthless epizoochorous diaspores from the dry tropical and subtropical semi-deserts, savannahs and grasslands of America, Africa and Madagascar. Adapted to stick to the feet rather than to the fur of grazing animals, these tough indehiscent or lately dehiscent fruits have sharply pointed spines, claws or horns. In the New World the most notorious of these fruits are the so-called devil's claws. The North American devil's claws belong to species of the genus *Proboscidea* (esp. *Proboscidea louisianica*) and their smaller relative *Martynia annua*, both members of the unicorn family (Martyniaceae). In South America it is their carnivorous relatives of the genus *Ibicella* that produce similar devil's claws, or unicorn fruits as they are sometimes called (e.g. *Ibicella lutea*). All have harmless-looking green fruits which reveal their true nature only after their fleshy outer part has withered away. As the exposed endocarps dry out, their elongated beak splits down the middle to produce a pair of sharply pointed, curved spurs transforming the diaspore into a vicious contraption, poised to cling to the hoof and bore into the skin.

The Old World members of the sesame family (Pedaliaceae) reveal their close relationship with the Martyniaceae by sharing the same ruthless means of dispersal and producing even nastier traps. As the author can testify from his own experience, the fruits of the Malagasy genus *Uncarina* are undoubtedly the most tenacious fruits of all epizoochoric diaspores. With their radiating spines crowned by a pair of sharply pointed, curved hooks, they rip into skin with ease and are impossible to remove from clothes without secateurs. The most infamous member of the sesame family is *Harpagophytum procumbens*, aptly called grappling hook, grapple plant, or, like its New World relatives, devil's claw. Used as a mouse trap in Madagascar, its preposterously vicious, woody pods can inflict terrible wounds on animals with cleft hoofs or relatively soft soles. Other members of the Pedaliaceae produce slightly more merciful but nonetheless pain-inflicting instruments of torture: the two sharp vertical spines on the upper side of the very hard, disc-shaped fruits of the African devil's thorn (*Dicerocaryum eriocarpum*) are thrust deep into the flesh of anyone who has the misfortune to tread on them.

Dispersal through human influences

Any kind of human-mediated dispersal is called *anthropochory*. This includes the seed-gathering activities of indigenous people following traditional lifestyles, as well as modern agricultural activities. Plants have always been moved from country to country and utilised away from their place of origin. Nowadays, farm animals are also moved over large distances

below: *Proboscidea louisianica* (Martyniaceae) – pale devil's claw; native to the southern United States – fruit with sharply pointed claws poised to cling to the feet of grazing animals. When immature, the green fruits are harmless and resemble a basally-swollen, curved bean pod. As the exposed endocarp dries out, its elongated beak splits down the middle into a pair of sharply pointed, curved spurs, revealing the cruel nature of the fruit; 17cm long

opposite: *Harpagophytum procumbens* (Pedaliaceae) – devil's claw, grapple plant; native to southern Africa and Madagascar – fruit with large woody grapples adapted to cling to the feet and fur of animals. The tough feet of ostriches are well protected against its sharp spines but animals with cleft hoofs or relatively soft soles can suffer terrible wounds; the tuberous root of devil's claw has been used by the Khoisan peoples of the Kalahari Desert for thousands of years to treat pain during pregnancy and to prepare ointments to heal sores, boils and other skin problems. Introduced into Europe in the early 1900s, extracts from dried roots are today sold as a natural remedy to relieve pain and inflammation in people with arthritis and other painful ailments; fruit 9cm long

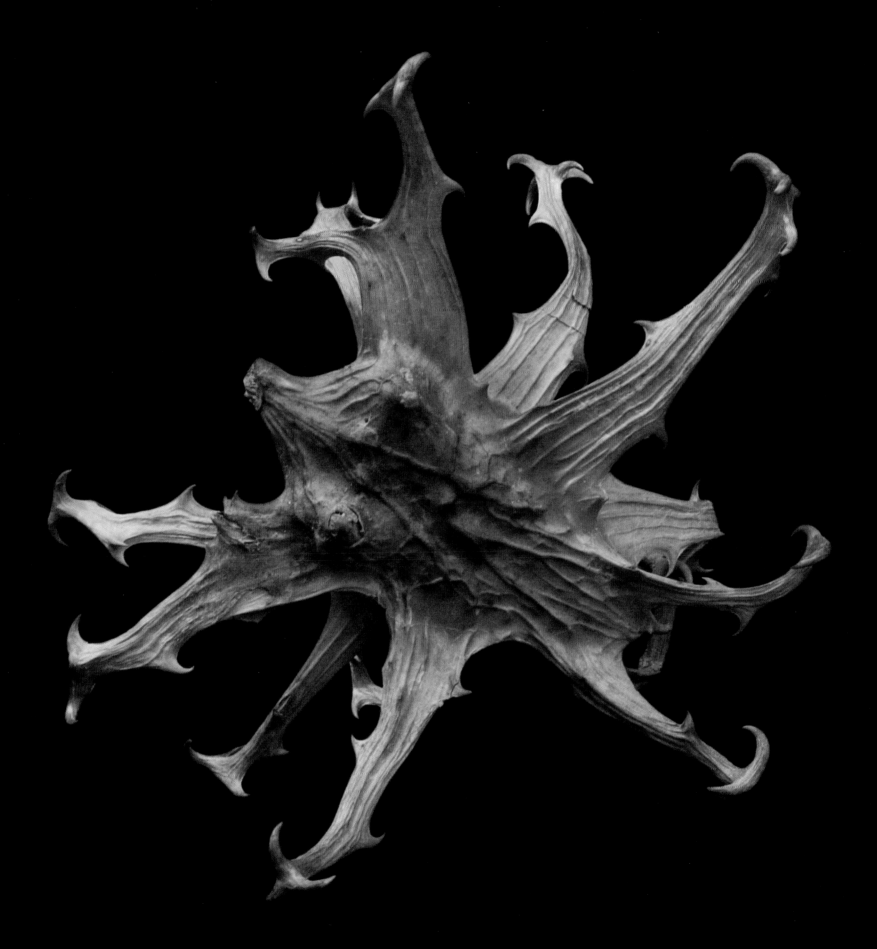

hundreds of kilometres. Humans also act as dispersers when they spit out a cherry stone, throw an apple core out of the car window or remove sticky hitchhikers from their clothes after a country ramble. Long-distance dispersal is provided by our enthusiasm for holidays in exotic places. Backpackers may travel the world wearing the same muddy hiking boots, unknowingly carrying a cornucopia of small seeds and other diaspores across biogeographical boundaries. A species suddenly taken to a new environment usually fails to survive but occasionally thrives and, in extreme cases, becomes invasive. In New Zealand, aliens such as gorse (*Ulex europaeus*, Fabaceae), blackberry (*Rubus fruticosus*, Rosaceae) from Europe and kahili ginger (*Hedychium gardnerianum*, Zingiberaceae) from South Africa take over natural habitats. In Britain, the number of aggressive weeds introduced as ornamental garden plants is increasing; they include giant hogweed (*Heracleum mentagazzianum*, Apiaceae) and Japanese knotweed (*Fallopia japonica*, Polygonaceae).

The impact of the accidental or intentional introduction of alien species into natural habitats should not be underestimated. Over a few hundred years, invasive species (both animals and plants) introduced by humans have become a major cause of extinction of native species in countries all over the world. The aesthetic qualities of the landscape may also be impaired by the newcomers.

Diaspores with no obvious adaptations for dispersal

Many small seeds shed from capsular fruits do not display any obvious adaptations to a particular mode of dispersal. Dust and balloon seeds typically weigh less than 0.005mg but small granular seeds under 0.05mg are also easily dispersed by the wind, even with no special adaptations (e.g. *Calluna vulgaris, Erica* spp., *Meconopsis cambrica, Papaver* spp.). Whether or not they are light enough to be carried by the wind, small granular seeds and other diaspores (e.g. the nutlets of the Lamiaceae) may be dispersed further after their liberation from the fruit by haphazard or accidental events. Rainwater can wash them away and they may even end up in a stream, where they float before being deposited on the bank or on a flood plain. Accidental endozoochoric dispersal is also a possible option. Many small, unspecialised diaspores are unintentionally ingested by grazing animals and pass through their gut largely unharmed, especially if they are rounded and compact. Mixed with mud, small diaspores can also stick to the feet, fur or feathers of animals who may carry them for a considerable distance. Surprisingly, many small seeds display conspicuous surface patterns which apparently have no significance for their dispersal. The magnificently ornate seeds of the Caryophyllaceae have already been mentioned. Similar seeds with intricate surface patterns are found in the related stone plant family (Aizoaceae), purslane family

opposite: *Uncarina* sp. (Pedaliaceae) – native to Madagascar – the fruits of *Uncarina*, probably the most tenacious of all fruits, are adapted to attach themselves to the fur or feathers of animals and birds. Once caught by the extremely sharp, curved hooks that crown the tip of the fruit's long radiating spines (see detail below) it is almost impossible to escape without injury; fruit 8cm in diameter

families. For example, the surface pattern of the spiny seeds of the monocotyledonous yellow star-of-Bethlehem (*Ornithogalum dubium*, Hyacinthaceae) is remarkably similar to the typical pattern of the dicotyledonous Caryophyllaceae (seed coat cells with undulating radial walls and finger-like projections).

Reticulate surface patterns similar to those found in dust and balloon seeds are also very common in larger seeds not dispersed by the wind, of a variety of unrelated families, e.g. *Ornithogalum nutans* (Hyacinthaceae), yellow-wort (*Blackstonia perfoliata*, Gentianaceae), common centaury (*Centaurium erythraea*, Gentianaceae), St. Helena boxwood (*Mellissia begonifolia*, Solanaceae), *Pholistoma auritum* (Boraginaceae) and *Eucalyptus* spp. (Myrtaceae). It is hard to believe that seeds would develop such elaborate surface structures if they did not provide an evolutionary advantage. Possible adaptational explanations may be found in the development of a seed or in details of its dispersal and germination ecology. Seemingly functionless structural peculiarities may be rudiments of long obsolete adaptations, their origins lost in the unknown evolutionary history of the plant. Nature does not always reveal all her secrets. Sometimes it may be sufficient that a character has no disadvantage for it to evolve and prevail.

Travellers in time and space

Seeds do not only travel in space but also in time. Their journey begins when they are shed by the parent plant. Whether they are taken far away from their place of origin by wind, water or animals or stay more or less in the same location, eventually all seeds touch ground at their ultimate destination. In the constantly warm, humid conditions of the tropical rainforests many seeds germinate immediately, otherwise they will dry out and die. Such desiccation-intolerant or *recalcitrant* seeds maintain a high water content and never slow their metabolic activity. Even if stored in moist conditions, recalcitrant seeds quickly lose their viability, a critical issue in trade and conservation. Around fifty per cent of tropical trees have desiccation-intolerant seeds, including crops such as durian (*Durio zibethinus*, Malvaceae), jackfruit (*Artocarpus heterophyllus*, Moraceae), rubber (*Hevea brasiliensis*, Euphorbiaceae) and cocoa (*Theobroma cacao*, Malvaceae). But plants with recalcitrant seeds are not restricted to the tropics. Many temperate broadleaved trees also have large recalcitrant seeds. Examples include oaks (*Quercus* spp., Fagaceae), sweet chestnut (*Castanea sativa*, Fagaceae) and horse chestnut (*Aesculus* spp., Sapindaceae). However, more than ninety per cent of all plants in temperate and dry climates have desiccation tolerant or *orthodox* seeds. In fact, losing water

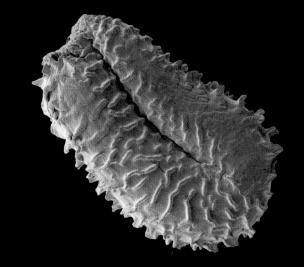

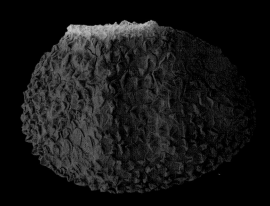

Seeds with no obvious adaptations to a specific mode of dispersal: *Damasonium alisma* (star fruit, Alismataceae; collected in the UK; 2.2mm long); *Anagallis arvensis* (scarlet pimpernel, Primulaceae; collected in the UK; 1.2mm in diameter); *Mammillaria theresae* (Cactaceae; native to Mexico; 1.5mm long); *Heuchera rubescens* (pink alumroot, Saxifragaceae, collected in Utah, USA); *Hermannia muricata* (Malvaceae, collected in South Africa; 1.3mm long).

seasonal and arid environments where orthodox seeds can rest in a dry, inactive state until the conditions for germination are favourable – the following spring perhaps, or the next rainy season. The quiescent embryo can sometimes wait for the auspicious moment for decades or even centuries within the safety of the seed coat or hardened fruit wall.

Older than Methusaleh

Legends, tales and rigorous scientific observation tell of seeds germinating after hundreds of years in a dry or frozen state arousing great public excitement and curiosity. Unfortunately, most of these claims arise in the imagination of the optimistic discoverer rather than accurate datings. In the nineteenth and early twentieth century it was reported that grains of wheat and barley found in three to six thousand year old Egyptian graves had germinated. Many respectable scientists, especially in the first half of the nineteenth century, supported the idea that such ancient seeds might still be viable. At that time the public was obsessed with ancient Egypt and fascinated by the idea that *mummy seeds* from the tombs of kings would spring to life after thousands of years. There is no scientific proof that these ancient cereal grains retained their viability but the tale of the *mummy seeds* is deeply entrenched in popular thinking. Modern research indicates that, under the conditions in which they were found, none of the species involved has seeds that could survive longer than a few decades. A much more credible discovery was made in a tomb in Santa Rosa de Tastil in Argentina. A viable seed of Indian shot (*Canna indica*, Cannaceae) was found enclosed in a shell of the Argentine walnut (*Juglans australis*, Juglandaceae) that was part of a rattle necklace. Radiocarbon-dating of surrounding charcoal remains and the nutshell in which the Canna seed must have been inserted while still soft puts the age of the necklace at about six hundred years.

The most extraordinary claim for a time-travelling diaspore was made in 1967 for seeds of the arctic lupin (*Lupinus arcticus*, Fabaceae) discovered by a mining operative three to six metres below ground in the frozen silt of Miller Creek in the Canadian Yukon Territory. They were thought to be about ten thousand years old, their longevity explained by their hard seed coat and the fact that they were deeply buried in permafrost. Unfortunately, their age was inferred from highly circumstantial evidence: they were found in an ancient rodent burrow associated with the skull of a collared lemming and the dating was suggested solely by analogy with similar rodent remains in Central Alaska. In the absence of any direct dating of the seeds by precise radiocarbon techniques the claim for the arctic lupin is open to considerable doubt and unlikely to be accepted by scientists.

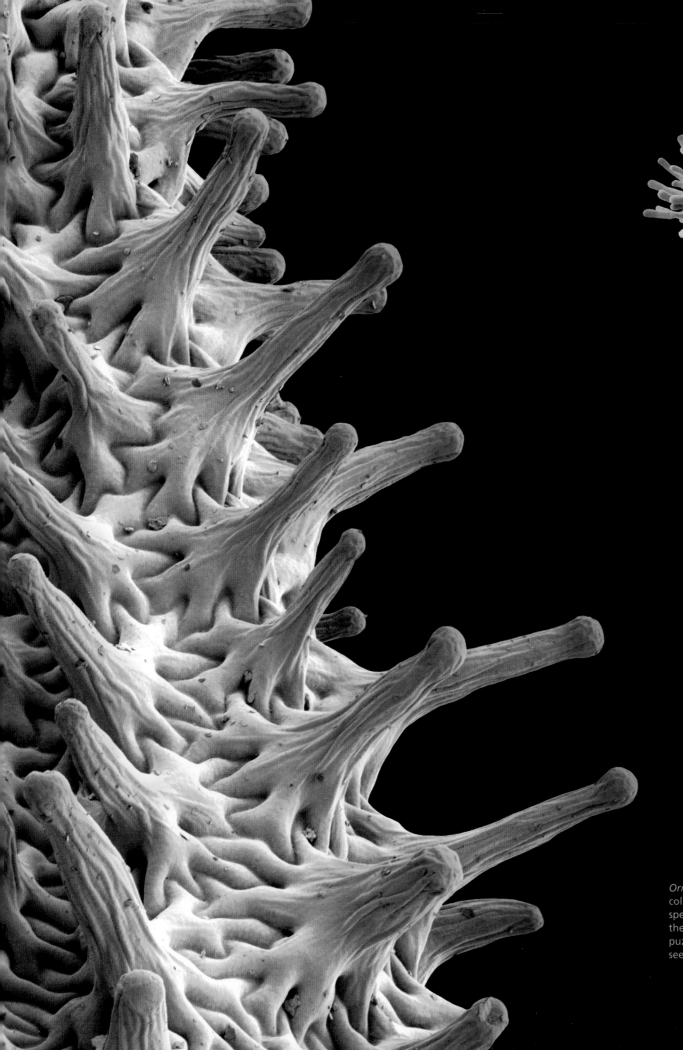
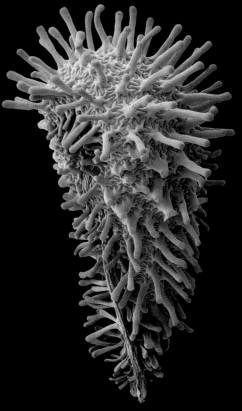

Ornithogalum dubium (Hyacinthaceae) – yellow star-of-Bethlehem; collected in South Africa – seed with no obvious adaptations to a specific mode of dispersal but displaying an intricate surface pattern: the cells of the seed coat are interlocked like the pieces of a jigsaw puzzle. Dispersal probably occurs when the capsules fling out the seeds as they sway in the wind; seed 1.1mm long

Seeds: Time Capsules of Life

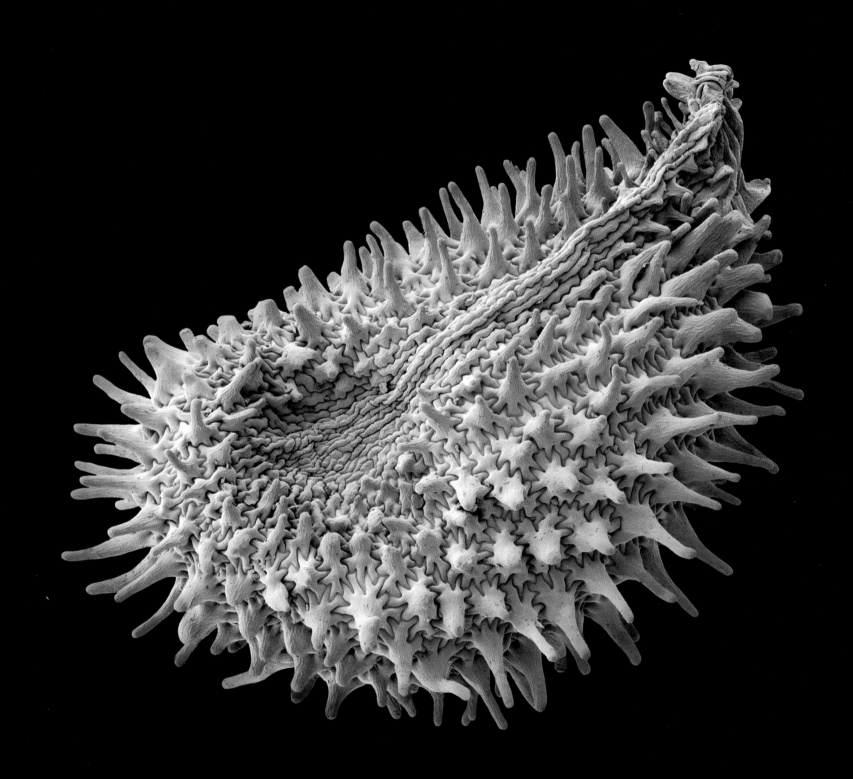

The oldest living seed whose exact age could be established belongs to the sacred lotus (*Nelumbo nucifera*, Nelumbonaceae). The sacred lotus has deep religious significance for Hindus and Buddhists and is known to have been cultivated in China for more than 3,000 years. It has long been claimed that its seeds – botanically nutlets of a multiple fruit with a very hard air- and water-impervious pericarp – can live for centuries. Scientific proof of their fabled longevity was provided in 1995 by Jane Shen-Miller who was able to germinate lotus seeds recovered from a dry lake bed in the former Manchuria (now north-eastern China). Modern accelerator mass spectroscopy techniques allowed precise radiocarbon-dating of a minute piece of the nutlets' thick, hard pericarp without killing the seeds. Using this method, it was proven that the oldest germinating seed was 1,288 (± 250) years.

Most recently, Israeli researchers have put forward a claim which, if proven, would put *Nelumbo nucifera* into second place among the longest living seeds. On June 12, 2005 the news about the germination of a 2,000 year-old seed of the date palm (*Phoenix dactylifera*, Palmae) hit the headlines of the *New York Times* and other newspapers. The seed belongs to a lot found in the 1970s during an archaeological excavation at Masada, the mountain fortress built by King Herod and famously occupied in 73AD by 960 Jewish rebels, who eventually committed mass suicide rather than surrender to the Romans. According to a radiocarbon dating carried out in Switzerland with a tiny piece of the germinated seed, its origin dates back 1,990 (± 50) years, that is to some time between 35BC and 65AD, just before the famous siege. The Israeli researchers have high hopes for this precious seedling, which they named *Methuselah* after the biblical figure who supposedly lived for 969 years. The native date palms of Judea were destroyed long ago; those grown there today originated in Iraq. The original Judean date was highly prized in antiquity for its medicinal and aphrodisiac qualities. However, date palms are dioecious and if *Methuselah* lives to adulthood, it will be about thirty years before it flowers and reveals its gender. Whether or not *Methuselah* turns out to be female and produces fruits, it may still possess the precious genetic traits of the original Judean date and become a treasure chest for breeders of date palms and medical researchers.

One year's seeds, seven years' weeds

Although the seeds of only a few species can survive such biblical time spans, most orthodox seeds are able to retain their viability for several decades in soil. Germination can be delayed by the absence of sufficient humidity or it can be suppressed by low temperatures and the lack of sunlight – for example, when seeds are buried. This makes sense since a small seedling germinating a foot underground would not stand a chance of ever reaching vital

below: *Nelumbo nucifera* (Nelumbonaceae) – sacred lotus; native from Asia to Australia – close-up of a flower showing the unusual (apocarpous) gynoecium, which consists of 12-40 separate carpels individually sunken into a large spongy floral axis. With its waterlily-like flowers and aquatic lifestyle the sacred lotus superficially resembles waterlilies (Nymphaeaceae) although its closest living relatives have been shown to be plane trees (Platanaceae) and members of the Proteaceae family. The sacred lotus has a deep religious meaning for Hindus and Buddhists in India, Tibet and China, where it has been cultivated since the twelfth century BC

opposite: *Nelumbo nucifera* (Nelumbonaceae) – sacred lotus; native from Asia to Australia – fruit consisting of the enlarged floral axis with numerous chambers, each containing a single-seeded nutlet. As the long flexible fruit stalks sway in the wind, the nutlets are ejected into water, where they immediately sink to the bottom. Lotus seeds are enclosed in the extremely hard pericarp of the nutlet and can retain their viability for more than a thousand years

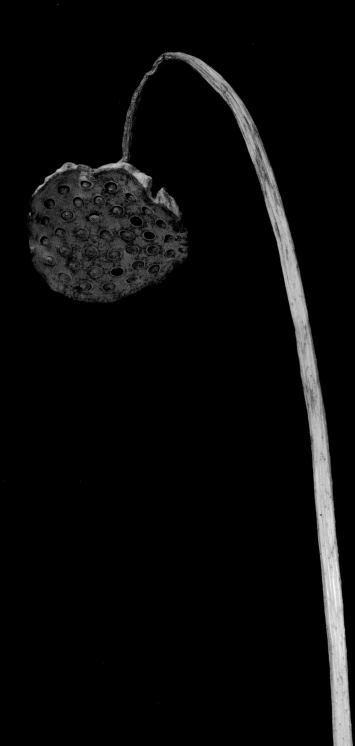

sunlight. Over the years, viable seeds accumulate on and in the ground to form a *soil seed bank*. The sudden appearance of poppies on the battlefields of World War I in Flanders is a good example. The churning up of the soil by the impact of shells and the digging of trenches and graves is believed to have brought to the surface the seeds of poppies long buried in the former wheatfields. It is not only the lack of water and sunlight that can delay germination for an infinite period of time. Many plants developed specific mechanisms which ensure that their seeds do not germinate immediately, even under the most favourable conditions. Seeds displaying this behaviour are called *dormant* and the various mechanisms which keep them asleep are summarized under the phenomenon of *dormancy*. There are three fundamentally different types of seed dormancy – physical, morphological and physiological.

The seed may remain dormant for a very simple reason like the presence of a hard, impermeable seed coat or pericarp, which prevents water from reaching the embryo inside. The hard seeds of many legumes and the nutlets of the sacred lotus are good examples of *physically dormant* diaspores, which often possess pre-formed openings that respond to high temperatures, large fluctuations in diurnal temperatures, or fire. Physically dormant seeds do not rely on accidental damage but on their highly specialised signal-detecting system to determine when to come to life. Other seeds fail to germinate immediately because their embryos are immature at the time of dispersal and the embryo has to undergo further growth and differentiation, for example those of many Ranunculaceae (e.g. *Anemone, Ranunculus*) and *Ginkgo biloba*. Although generally referred to as *morphologically dormant*, because of their high water content and metabolic activity (the growing embryo) it is debatable whether these seeds should be regarded as truly dormant. As any gardener can testify, the most complicated reason for delayed germination is physiological dormancy. With no visible indicators and a multitude of different causes, physiologically dormant seeds germinate only when certain chemical changes have occurred in them. Environmental triggers for the changes can be low or high temperatures, smoke or fire. To complicate matters further, different types of dormancy can be combined in the same seed, either morphological and physiological dormancy, or, rarely, physical and physiological dormancy. The most important difference is that physiological dormancy is reversible whereas physical and morphological dormancy are not. The advantage of physiological dormancy is its flexibility. Some arable weeds go through an annual dormancy cycle enabling them to germinate only at a certain time of year.

The displacement of seeds underground and dormancy add to the creation of a persistent soil seed bank. The presence of many annual weeds in soil seed banks and

differences in the depth of dormancy between seeds of the same fruit explain the popular saying "one year's seeds, seven years' weeds". If weeds are left to disperse their seeds for just one season, their offspring will haunt the careless gardener for many years to come. Of course, dormant seeds did not evolve to escape the watchful eye of the proud gardener. The real strategy behind dormancy is to achieve timely dispersal. For a species to survive, it is absolutely essential that at least some of its seeds germinate in the right place at the right time. Dormancy is a mechanism that supports this goal in several ways: it allows more time for the various dispersal agents to carry the seeds over longer distances and, most importantly, determines the best time for germination. In a temperate climate many seeds germinate in spring to make the most of the growing season and ensure that they are well established before winter. Others are programmed to germinate in autumn so that they develop a strong root system to protect them against drought in summer. In addition, the different depths of dormancy in seeds of the same plant ensure that germination is spread over time, reducing the risk of losing an entire generation through some catastrophic event such as fire, drought or frost.

Time capsules of life

The ability of orthodox seeds to survive for a long time in a dry state is their most astonishing and momentous quality. Soil seed banks and the aerial seed banks created by serotinous species like the jack pine (*Pinus banksiana*) and many South African and Australian Proteaceae provide life insurance for the species that invest in them. For all annual herbs and many perennial plants in fire-prone habitats, their naturally created seed cache is the only way they can survive adverse environmental conditions. Natural seed banks therefore play a major role in plant succession and the evolution of plant communities.

The extraordinary survival skills of seeds also had a great influence on the evolution of our civilisation. The development of modern societies is strongly linked to advances in agriculture. Agriculture and permanent settlements became possible only when people discovered that they could collect and store seeds to cultivate their own supply of a certain crop rather than having to lead a nomadic hunter-gatherer existence. In the 1960s people began to realise that modern high-yielding crop varieties were displacing the tremendous diversity of the local crop cultivars that indigenous people all over the world had been selecting for centuries. Traditional local varieties are not only better adapted to the climatic and edaphic conditions in their place of origin, they also carry valuable genetic traits useful in the breeding of new varieties. Their small size and longevity make seeds an extremely efficient means of preserving precious botanical germplasm.

Nemesia versicolor (Plantaginaceae) – leeubekkie (Afrikaans); collected in South Africa (Eastern Cape) – seed with peripheral wing to assist wind dispersal; seed (including wing) 2.4mm in diameter

Mature orthodox seeds can be dried without damage to very low levels of moisture content (1 to 5 per cent). As their moisture content drops, they become increasingly tolerant of cold temperatures. The explanation for this phenomenon lies in the physical qualities of water. In plant cells, water can be either free (*free water*) or bound to molecules (*bound water*) such as proteins, sugars and polysaccharides. At temperatures below 0°C, free water freezes and expands while forming ice crystals. As the ice crystals grow they damage the cell organels and perforate cell membranes and walls, literally stabbing the cells to death. Bound water, with molecules that are electromagnetically attached to other, larger molecules is unable to produce lethal ice crystals. Recalcitrant seeds are rich in free water and so cannot survive sub-zero temperatures whereas orthodox seeds, when properly dried, contain almost no free water. It was established long ago that the longevity of mature orthodox seeds in hermetic storage increases as the moisture content of embryo and endosperm drops and the ambient temperature is lowered. This predictable and quantifiable relationship between water content and storage temperature is summarized in Harrington's 1973 rule of thumb. It predicts a doubling of storage life for every one per cent reduction in moisture content (based on the fresh weight), and a doubling of storage life for every 10°C reduction in storage temperature. Although almost impossible to verify experimentally, this simple rule forms the theoretical basis of all modern seed storage. Predictions suggest that seed longevity under conventional seed bank conditions (storage in air-tight containers at around minus 20°C) varies from a few decades for the shortest-lived species to over a thousand years for the longest-lived.

The first application of dedicated storage facilities for seeds, called *seed banks*, was to preserve varieties of major crops, especially cereals. Nowadays, the continuing destruction of our environment brought about by the increasing world population has created a new challenge for seed banks: the preservation of seeds of wild species which have been driven to the brink of extinction in their natural habitats. Since 1600 there have been 654 documented extinctions of plant species but this is undoubtedly a massive underestimate of the true number. The mass extinction caused by the destruction of tropical rainforests has been likened to the fall of the dinosaurs. With vast areas of our planet still botanically unexplored, there is no way of knowing how many plant species are currently threatened with extinction. Estimates vary between 30,000 and 100,000. With so many species facing being wiped off the planet for ever, seed banks have acquired a unique role: they provide a man-made life insurance for threatened plants and transform ordinary seeds into true time capsules of life.

Parnassia fimbriata (Parnassiaceae) – fringed grass-of-Parnassus; collected in Oregon, USA – seed with loose, bag-like seed coat displaying the typical honeycomb pattern of wind-dispersed balloon seeds; 1.3mm long

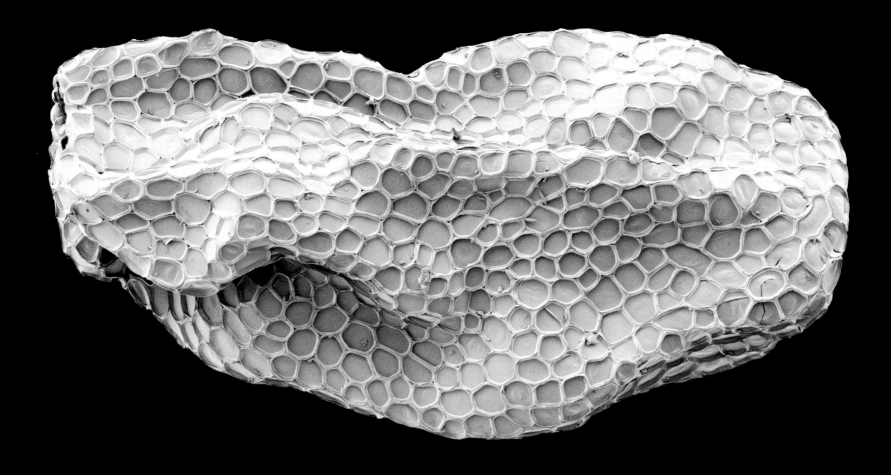

THE MILLENNIUM SEED BANK

Noah's Ark of the 21st Century and Beyond

The most ambitious banking enterprise entirely dedicated to storing the seeds of wild plant species is the Millennium Seed Bank Project (MSBP) managed by the Royal Botanic Gardens, Kew. An international conservation project established to mark the new millennium, the MSBP was founded in 2000 with funding from the Millennium Commission, the Wellcome Trust, Orange plc, and other corporate and private sponsors. The Royal Botanic Gardens, Kew, has operated a seed bank for wild species since the early 1970s. At the start of the MSB Project, it moved into the new award-winning Wellcome Trust Millennium Building designed by Stanton Williams Architects. Located in the grounds of Wakehurst Place in Sussex, Kew's "country garden", it offers a world-class facility located in an Area of Outstanding Natural Beauty. It has space to store thousands of seed samples in an underground vault, advanced seed research and processing facilities, and public display areas.

By 2010, the MSB Project aims to have collected, conserved and researched 10 per cent of the world's flora – approximately 24,000 species (based on Mabberley's 1987 conservative estimate of 242,000 species of seed plants worldwide). Kew's conservationists are focussing their efforts on endangered, endemic and locally important economic species in arid and semi-arid regions. Concentrating on plants from areas with seasonal or erratic rainfall has clear practical benefits for seed banking. The percentage of species with orthodox seeds (which can be banked successfully) is higher than in wetter habitats such as tropical rainforests, and the clear seasonality of vegetation processes in drylands permits more predictable timing of seed-collecting activities than is possible in warm, humid climates. Another practical advantage is that most species in drylands are herbs, shrubs or small trees with readily accessible seeds, making collection easier. More important than such practical considerations is the fact that drylands account for a third of the Earth's surface and include many of the world's poorest countries. Although they receive less public attention than tropical rainforests, drylands support a fifth of the world's population and sustain a

opposite: Public entrance of the Millennium Seed Bank at Wakehurst Place in Sussex. Managed by the Seed Conservation Department of the Royal Botanic Gardens, Kew, the building houses one of the United Kingdom's most ambitious conservation initiatives: by the year 2010 the Millennium Seed Bank Project aims to have collected and preserved the seeds of approximately 24,000 wild plant species from all over the world

above: Seeds germinating in a petri-dish with water-agar during a germination test at the Millennium Seed Bank

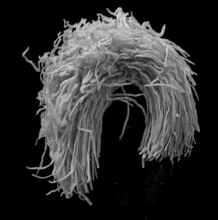

tremendous diversity of plant life. Drylands are among the most threatened environments with huge areas lost every year to progressive desertification, thus reducing the natural diversity of plants that are a source of livelihood to many, especially in poorer regions.

The huge task of collecting the seeds of almost 25,000 species is based on extensive international collaboration and information sharing. To date, forty-five partner institutions from eighteen countries have joined the Millennium Seed Bank Project; these are (in alphabetical order): Australia, Botswana, Burkina Faso, Chile, China, Egypt, Jordan, Kenya, Lebanon, Madagascar, Malawi, Mali, Mexico, Namibia, Saudi Arabia, South Africa, Tanzania and the USA. A substantial percentage of the seeds collected is kept in the country of origin if adequate storage facilities are available locally. Otherwise, half the collected seeds of each species are set aside and kept at the Millennium Seed Bank until respective partner countries have their own seed banks. The international collecting programme is carried out in accordance with national and international law, and is particularly cognisant of the Convention on Biological Diversity agreed at the 1992 Rio de Janeiro Earth Summit.

The actual seed banking procedure is quite straightforward. On arrival, seeds are cleaned to remove unnecessary fruit parts, and diseased, infested and empty seeds are discarded. If a collection is of acceptable quality, its size is estimated. The next step is to dry the seeds under constant conditions (15 per cent relative humidity at 15° C) for at least four weeks in a dedicated dry room. The controlled desiccation lowers the moisture content to around 5 per cent. They are then ready for storage at minus 20° C. To ensure that a collection is viable, a representative sample is tested for germinability, usually using petri-dishes with water-agar. Ideally, collections should have a viability of 75-80 per cent to guarantee minimal genetic changes. Dormant or dead seeds are identified by vital staining (e.g. tetrazolium). This test usually takes place after at least one week of cold storage. During the long-term storage of the seeds, germination tests are carried out at regular ten-year intervals to monitor the quality of the collections.

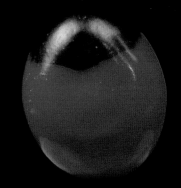

above: *Acacia cyclops* (Fabaceae) – coastal wattle; native to south-western Australia – seed surrounded by a bright orange-coloured aril formed by a double layer of the funiculus that encircles the seed in one direction and then folds back to surround it once more in the opposite direction; 9mm long (including aril)

centre: *Abrus precatorius* (Fabaceae) – rosary pea; found in all tropical regions – hard, shiny seed mimicking a fleshy seed coat or a berry. Birds are tricked into thinking the seeds are edible and move them over short distances until they discover the scam. Rosary peas contain Abrin, a ribosome-inactivating protein and one of the most deadly plant toxins known (0.5g are fatal in humans); 5.5mm long

top: *Strelitzia reginae* (Strelitziaceae) – bird-of-paradise flower; native to South Africa – seed with a conspicuous aril of thick, bright orange-red hairs, which is very attractive to birds; c.1cm long (including aril)

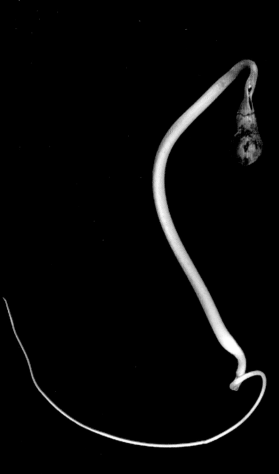

above: *Euphorbia damarana* (Euphorbiaceae) – milkbush; collected in Namibia – seedling; this shrubby succulent *Euphorbia* species is one of the most valuable plants in the southern African deserts. Although poisonous to humans, it is a favourite food plant of rhinos, kudus, springbok and gemsbok, and provides shelter for small mammals

opposite top left: Seed containers ready to be sealed for cold storage

opposite top right: Cut-test: a quick and easy way to establish the quality of a collection; amount, colour and texture of endosperm and embryo allow conclusions as to the quality of a seed

top right: Conventional pickling jars have proved ideal to store seeds under dry conditions at minus 20°C

top far right: Cold-room with space-efficient compacter storage units at the Millennium Seed Bank

The United Kingdom Programme has already collected seeds from over 96 per cent of native higher plants. This is the first time that any country has underpinned the conservation of its flora in this way. The international drylands programme is well under way and accelerating as training is complemented by experience and hard work. To help conserve the seeds of difficult-to-store species, Kew's Seed Conservation Department runs an active research programme in fundamental aspects of seed storage, longevity and germination.

Seed collections of wild species are held for long periods to provide "start-up" stocks that will enable future generations to adapt to change. Safe in seed banks, species valuable for human well-being are preserved even if they become extinct in the wild. More importantly, they will be available to ensure the recovery of damaged ecosystems. Perhaps reflecting the early signs of humanity's adaptation to the changes it is bringing about in ecosystems, significant use of the Millennium Seed Bank's collections is already being made. Over the past five years, more than 3,400 collections have been made available to thirty-seven countries outside Europe, rich and poor in almost equal measure. Collections are distributed under terms which secure the rights of the country of origin to any ensuing benefits and have been requested to support work in all seven areas of human sustainability: agriculture, the atmosphere (in connection with carbon dioxide levels and climate change), biodiversity, chemicals, energy, health, and water.

Seed Banks are a highly cost effective way to preserve genetic variation in and between individual species. They occupy little space and require little attention for considerable periods. The total cost of the ten-year Millennium Seed Bank Project is £80 million, including the construction of the new building. This compares with the cost of administering a UK general election; forty conventional cruise missiles; a 10 pence insurance premium against the loss of plant diversity, upon which all life depends, levied on each of the additional 800 million people likely to join the human race in the coming decade.

Why risk so much for so little?

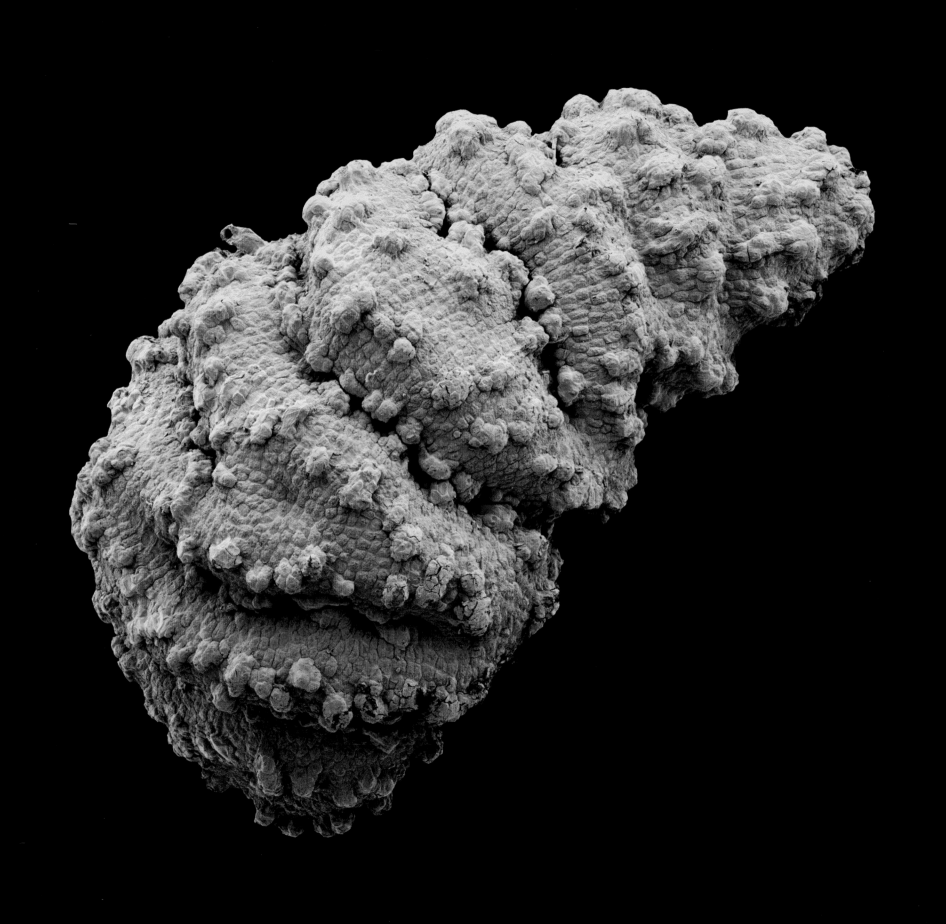

AN ARCHITECTURAL BLUEPRINT

ROB KESSELER

...science is what you know, art is what you do.
The best art is founded on the best science in any given manner

William R. Lethaby
Form in Civilization, 1922

Hermannia sp. (Malvaceae) – collected in South Africa, Limpopo Province – seed with no obvious adaptations to a specific mode of dispersal but displaying an interesting surface pattern; 1.4mm long

opposite: *Cleome* sp. (Capparaceae) – spiderflower; collected in Madagascar – seed; 1.5mm in diameter

top: Vaulted ceiling at Christ Church Cathedral, Oxford

above: Leaves of the palm *Licuala grandis* – Vanuatu fan palm – photographed in the Palm House at Kew

From buds to blobs

The relationship between the plant world and architecture is as old as the first rudimentary dwellings that humans constructed as they emerged from their caves; habitation out of vegetation. To call their structures primitive, grossly undervalues their understanding of the functional value of the material they used to create simple structures, an understanding based on observation and experimentation: the way columnar stem forms have great supporting properties or the ribbed, radiating surfaces of palm leaves make ideal roof coverings, providing shade and channelling water away from the centre. In this way the early builders emulated plant properties and characteristics in their constructions establishing a link between plants and form, creating an architectural precedent that is as relevant today as it was then.

The transition from primitive dwellings to temples of culture and commerce has seen much appropriation and many adaptations of botanical form and decoration. The Corinthian capital that preoccupied Vitruvius (c.70-25 BC) with its apparent origins in the Acanthus plant proved so popular that it has survived for over two and a half millennia. This passion for aerial inflorescence was brought to a logical conclusion on top of the pillars within Oxford University Museum, where Irish stonemasons under the tutelage of John Ruskin (1819-1900), carved directly from specimens from the University Botanic Gardens. Each capitol represented a different plant so that the architectural decoration shifted from a purely ornamental embellishment to one where the whole building could be read as a work of botanical reference.

Writing in 1922 on the relationship between science and architecture, William Lethaby, architect and first principal of the Central School of Arts & Crafts in London poignantly anticipated the dynamic changes that were to take place in architectural design during the twentieth century. Emerging from an era in which Ruskinian ideals and the naturalistic forms of the Arts and Crafts Movement had mutated into the more organically opulent Art Nouveau, the architecture of the transitional post-war era became subject to the twin influences of Constructivism and Expressionism that subsequently defined the territory for Modernism. Drawing upon the potential for constructing in new materials, reinforced concrete, steel and glass, engineers enabled architects to realise visionary forms that marked

a radical shift in style, scale and theoretical understanding in the vital role architecture plays in how we live and work. Simple housing projects, new factories and exhibition pavilions all offered opportunities for architects to redefine urban topographics.

With the onset of Modernism with its reductive aesthetic and purity of function there appeared to be little room for what was seen as superfluous ornament. Ironically the infamous commentary on "ornament as crime" by the Austrian architect Adolph Loos, (1870-1993), had its roots in the honesty of labour and craftsmanship that had been so clearly espoused by William Morris (1834-96) and the Arts and Crafts Movement. This was to prove a lean time for botanical reference within architecture, but as in all things, styles change, new technologies evolve and fresh opportunities present themselves. It was through a structural experimentation using principles of geometry and mathematics that pioneered a shift away from the planar geometry of Modernism led by the visionary engineer-architect-designer, Buckminster Fuller (1895-1983). His geodesic domes and the gridshell structures of later architects like Frei Otto (1925-) echo the spirit of the age of space exploration and move architecture into a global dimension. Their forms and structures have parallels within the very building blocks of nature, questioning how we want to live in the future. Their optimistic belief in technology was rewarded as emergent computer sciences evolved to the extent where complex bio-morphic forms could be conceived and visualised, calculated and constructed. The pace of this change has been swift, as has the generation of diversity of new forms and the languages to describe them: blobitechture, biological baroque, biotechnic, technorganic, biomorphism, organicism, bio-metaphor, organicity, evolutionary algorithms, quasi genetic coding schemes.

The relationship to natural form is evolving beyond its superficial trademark of swelling asymmetric forms, beyond what the biologist D'Arcy Thompson described as "form as a diagram of forces", in his influential book *On Growth and Form* (1917). Advances in understanding within the fields of bio-mechanics and botanical sciences coupled with the rapid development of responsive materials and computer simulations enable architects to *evolve* buildings as total living entities. Like their botanical and animal counterparts buildings are becoming more responsive to climatic conditions, and as the development of self-healing properties in materials becomes a real possibility the relationship between architecture and the plant world is evolving to a position of emulation rather than imitation.

Wherever the new technologies take us, on examining these minute seed structures with their astounding diversity of form, complex articulated surfaces and tensile membranes it is a clear reminder that nature continues to provide us with inspirational examples that challenge our own creativity.

opposite: *Encephalartos inopinus* (Zamiaceae); endemic to South Africa – male cones

above: 30 St Mary Axe, London. Foster & Partners, commonly called "The Gherkin", "Crystal Phallus", or "Swiss Re Tower" after its owner and principal occupant. The 180m tall building is famous for its unconventional architecture

Delphinium peregrinum (Ranunculaceae) – larkspur; native to the Mediterranean – wind-dispersed seed covered by papery lamellae helically arranged around the longitudinal axis; 1.2mm in diameter

opposite: Downland Gridshell, Weald and Downland Museum, West Sussex. Edward Cullinan Architects

opposite: Interior staircase of The Monument, London, built in 1671-79 by Sir Christopher Wren (1632-1723)

Prenia tetragona (Aizoaceae) – collected in South Africa, Cape Province – seed; as is typical in the stone plant family, the seed is shed from hygrochastically opening capsules; although it boasts conspicuous sculpturing; up to 1.3mm in diameter

Pholistoma auritum (Boraginaceae) – fiesta flower; collected in California, USA – seed; the globose shape and honeycomb pattern allow the seed to be blown along the ground by the wind or washed along by rainwater; 2.9mm in diameter

opposite: The Eden Project, Cornwall, Nicholas Grimshaw & Partners

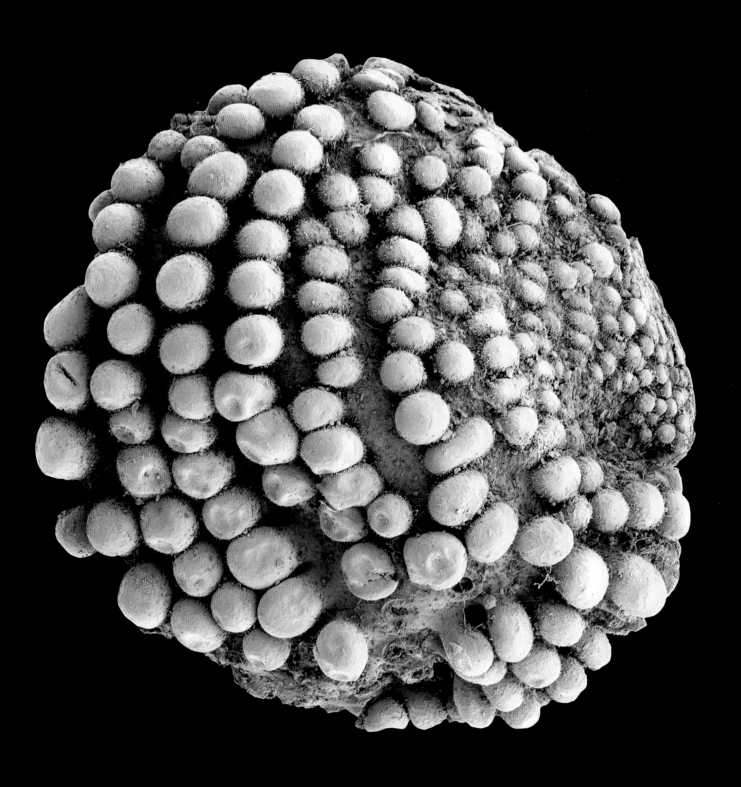

Glottiphyllum oligocarpum (Aizoaceae) – collected in South Africa, Western Cape – seed; as in other members of the stone plant family, the seeds of this species are shed from hygrochastically opening capsules; they have conspicuous sculpturing but show no obvious adaptations to a specific mode of dispersal; 1.2mm in diameter

opposite: Selfridges, Birmingham, Future Systems

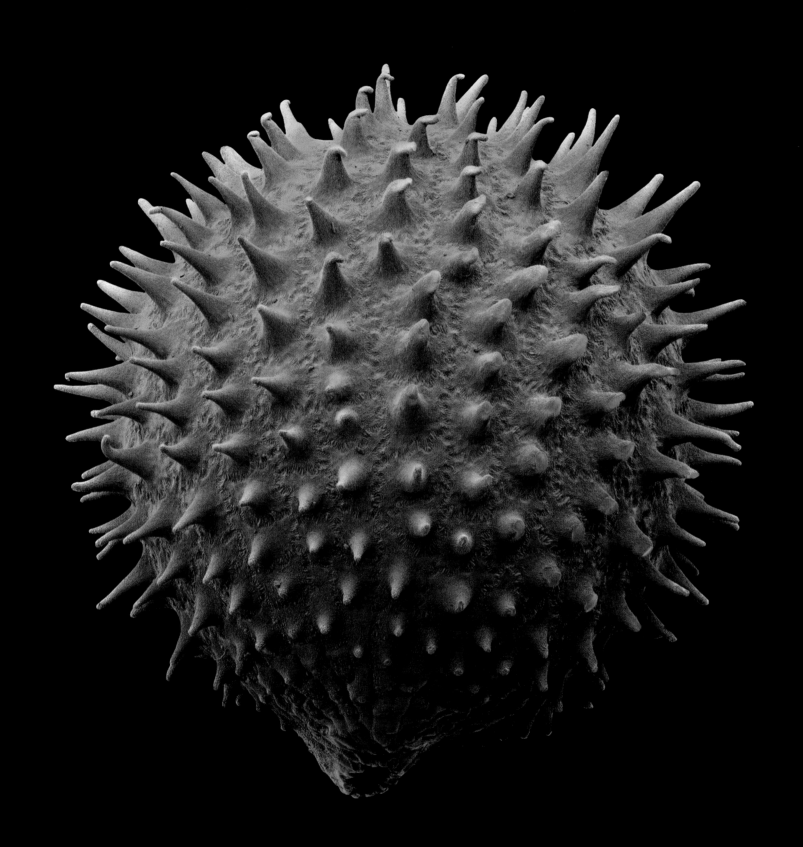

PHYTOPIA

ROB KESSELER

Tolmiea menziesii (Saxifragaceae) – collected in Oregon, USA – the spiny seed is shed from capsules and has no obvious adaptations to a specific mode of dispersal; however, its spiny surface may become entangled in the fur an animal, or wind or rainwater may drive the globular seed over the ground; 0.6mm in diameter

The difference between looking and seeing... an awesome clarity

*"Then you should say what you mean," the March Hare went on.
"I do," Alice hastily replied; "at least –
at least I mean what I say – that's the same thing, you know."
"Not the same thing a bit!" said the Hatter.
"Why, you might as well say that 'I see what I eat' is the same as 'I eat what I see'!"*

Lewis Carroll, *Alice's Adventures in Wonderland*

The urge to portray and understand the flowers and plants that surround us has a long and glorious history. They have become powerful symbols that carry many messages, markers with which we retain contact with the natural world; it is hard to imagine a part of our lives that they do not touch upon. As the systematic study of plants has evolved, so too have attitudes to picturing plants, which in their own way have propagated a complex and colourful genus of criticism, full of contradictory opinions and attitudes that reflect divergent attitudes within the respective fields of art and science:

In the art world illustration is a dirty word. It suggests slavish copying. It's seen as belonging to the world of functionality. And we all know art is at its best when it transcends functionality – when in short it is useless.[13]

Botanical illustrations have very little to do with art, but belong rather to the realm of the sciences. Aesthetic considerations are wholly inappropriate, and beauty is a pleasant but wholly irrelevant, side effect.[14]

Apart from doing a great disservice to the artists involved these two statements also seem to imply that the work only has validity within the immediate community for which it was created. In reality, the fabulous diversity of botanical art has been responsible for creating, inspiring and informing new audiences, reflecting the ideals and aspirations of the societies in which it was created. What these comments demonstrate is that every discipline has its taboos and agreed modes of operation beyond which consensus suggests we do not tread, even within an art world where pride is taken in subverting the rules.

The languages engendered by contemporary art and nature are complex and cyclical. However, whilst it is now recognised that our experience of nature is culturally mediated it is important not to lose sight of the object of that mediation. Discourse on the nature of

Stellaria pungens (Caryophyllaceae) – prickly starwort; collected in New South Wales, Australia – seed with no obvious adaptations to dispersal but displaying the intricate surface pattern typical of the pink family: the slightly papillose cells of the seed coat are interlocked like the pieces of a jigsaw puzzle; 1.5mm in diameter

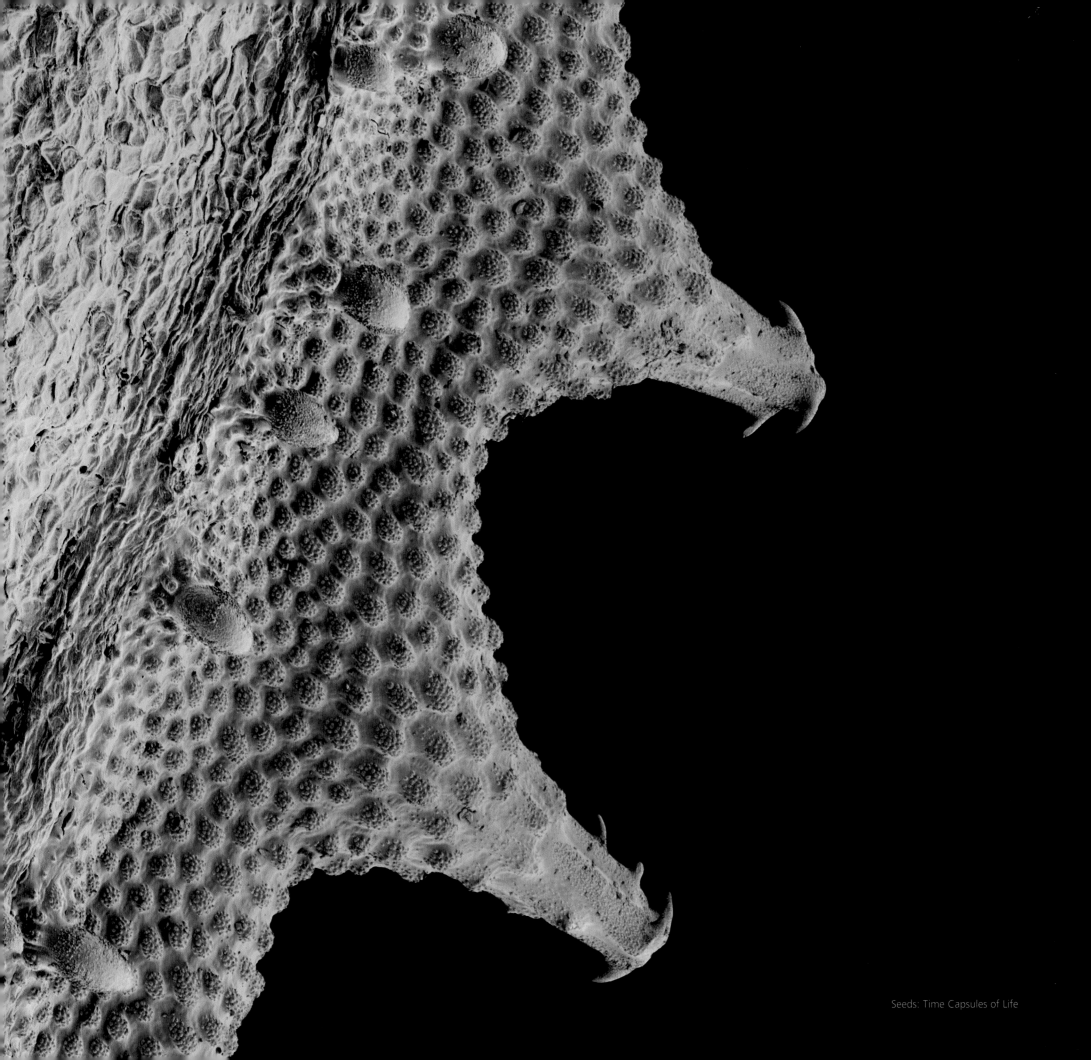

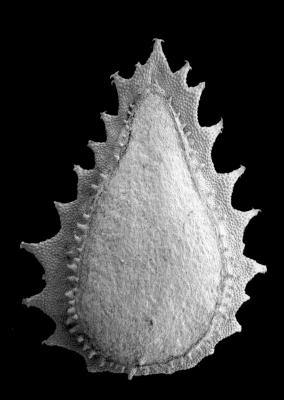

Trichodesma africanum (Boraginaceae) – collected in Saudi Arabia – single-seeded nutlet 3.9mm long (including spines)

opposite: detail of surface displaying the most common and effective strategy to ensure dispersal on an animal, namely to cover the surface of the diaspore with barbed spines or hooks. Like many members of the borage family, the ovary of *Trichodesma* is deeply four-lobed and splits at maturity into four single-seeded nutlets.

nature can be erosive, like a photocopy of a photocopy ; it can degenerate to the point where the image is there but the detail has gone. The difference between seeing what you eat and eating what you see is an important distinction. The human instinct to distinguish between the edible and the poisonous, an enemy and one's prey is an instinctive evolutionary tool essential for survival. However, the current pace of life, speed of change and diversity of the objects and images that pass before our eyes has evolved into overwhelming visual miasma, requiring us to become adept at instantly identifying, assimilating and cataloguing them. Have we now become expert at recognition at the expense of a more perceptive understanding and appreciation that arises from a concentrated examination of any given subject? Has society abrogated that responsibility to 'experts' with their rational taxonomic and genetic systems of identification and classification?

Take for example a common meadow buttercup (*Ranunculus acris*), easily recognisable across a field. The expert will tell us that the family to which it belongs (Ranunculaceae) is a primitive flowering plant not easily defined in evolutionary terms, with over thirty different wild varieties in the United Kingdom, including not only buttercups but also spearwort, hellebore, water-crowfoot, pasqueflower, wood anemone, columbine and traveller's-joy. The non-specialist in contrast, whilst its descriptive name might revive childhood memories of holding a flower up under the chin to detect a fondness for butter, might nevertheless be hard pressed to say how many petals it has.

In the creation of this book many hours have been spent examining the complexities of very small seeds, highly magnified on a scanning electron microscope (SEM). With such a tool the diversity of form and structural complexity of seeds is staggering; that such detail exists on such a minute scale is difficult to comprehend and one can only marvel at the technology that makes this possible. Returning to examine the plants and flowers from which the seeds had been collected necessitates a more concentrated inspection and in so doing one is made aware of the sophistication and power of our own in-built optical technologies. The difference between looking and seeing is thrown into sharp focus.

The creation of images for this book was conceived to revive the spirit of looking. The macro photography of the original flowers brings them into hyper realistic focus to encourage the reader to look again at the familiar flower, whether a roadside weed or florist's bouquet. Under the SEM the technology works its magic but presents us with a black and white image, which is subsequently coloured. This often provokes the question, "Is this the real colour of the seed?", to which the answer is no. And so how is the colour chosen and why?

Without going too deeply into the philosophical conundrum of what is colour, it is worth remembering that when we look at a flower we do not see it in the same way as an

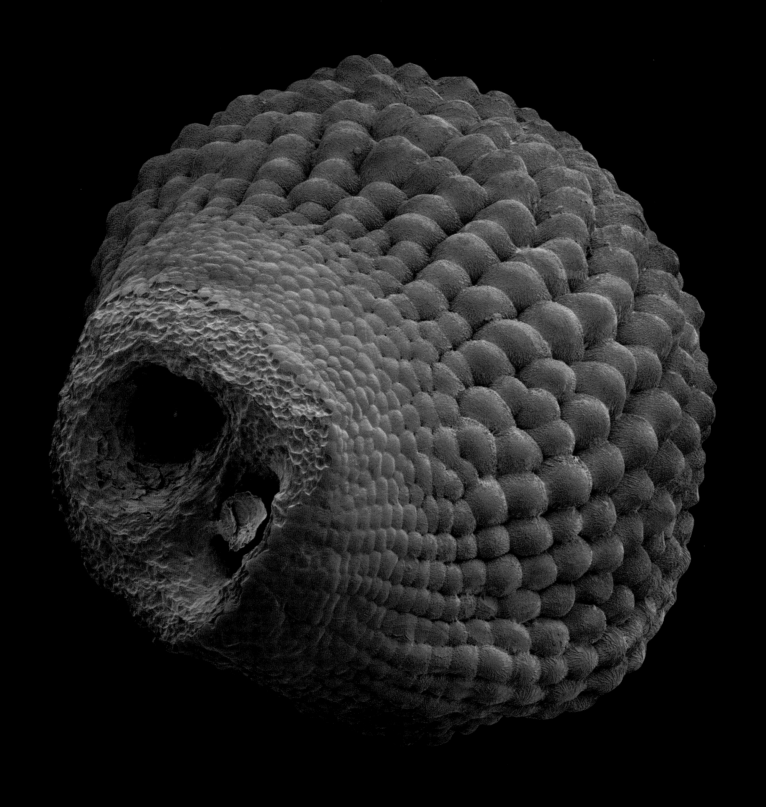

insect. Flowers have evolved complex strategies to ensure they attract the appropriate pollinators, through smell, morphological imitation, colour coding, and patterning. Most insects have greater sensitivity to colours at the blue end of the spectrum and are able to detect ultra-violet colours, revealing patterns that direct the insect to the pollen bearing parts of the flower like an aircraft guided to a safe landing by runway lights at night.

Working together as an artist and scientist with the same shared fascination for seeds, as for the plants from which we collected them, we too employed diverse strategies to ensure that our subject attracts as many "visitors" as possible. Under normal conditions scientific research is restricted to a very focussed methodological approach, but in this case we selected our samples with the express purpose of revealing extremities of form. Since bio-diversity is so vital within our ecosystems for the continued longevity of human existence we believe that it is important to celebrate this diversity. The selected specimens were composed and photographed to reveal their morphological characteristics with an intimate and awesome clarity. To these grey images colour has been added, a chromatic interference often inspired by the hues of the original flower, a subtle blending reminiscent of hand-tinted photogravures, lending the images a mysterious otherworldliness that transforms a spectator from one who just looks to one who sees and wants to know more. The colour guides the eye and moistens enquiry to stimulate what Mark Gisborn refers to as "imagination of resemblance"[15] through a total fusion of contemporary scientific and artistic practice. In so doing we hope to revive the importance of collaboration between artists and botanical scientists, an importance that was succinctly highlighted by Dr. T.J.Diffey in his essay, "Natural beauty without metaphysics":[16]

For art to continue this traditional task of making nature aesthetically accessible to a wider public, at least three things are necessary: first, nature requires mediation to an audience because that audience cannot appreciate it unaided; secondly, the art which mediates nature must not be relentlessly formal and abstract in its intentions; thirdly, nature must be available to the artist as a subject to study.

Echinocereus laui (Cactaceae) – Lau's hedgehog cactus; native to Mexico; berry-borne seed with no obvious adaptations to a specific mode of dispersal other than a rounded shape and rather smooth surface allowing easy passage through the gut; seed 1.43mm long

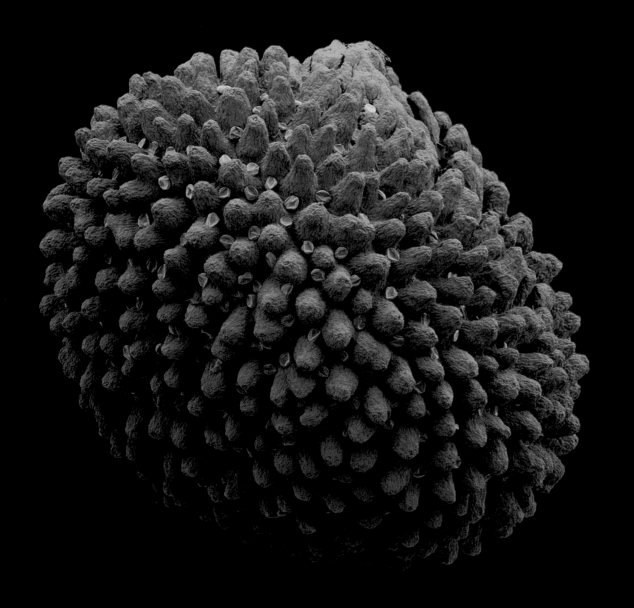
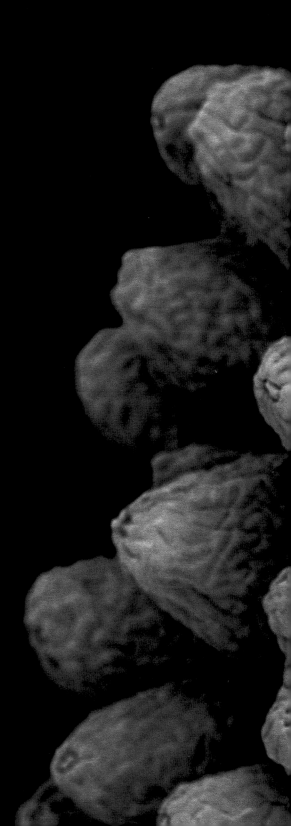

Scutellaria galericulata (Lamiaceae) – common skullcap; collected in the UK – single-seeded nutlet; a typical member of the mint family, this species has a deeply four-lobed ovary that splits at maturity into four single-seeded nutlets. The papillose surface is well equipped with oil glands (yellow); 1.6mm long

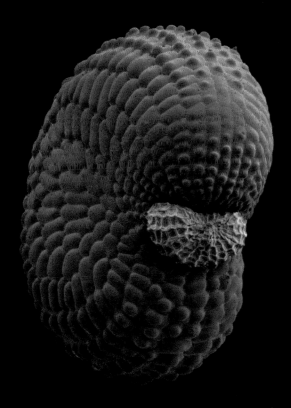
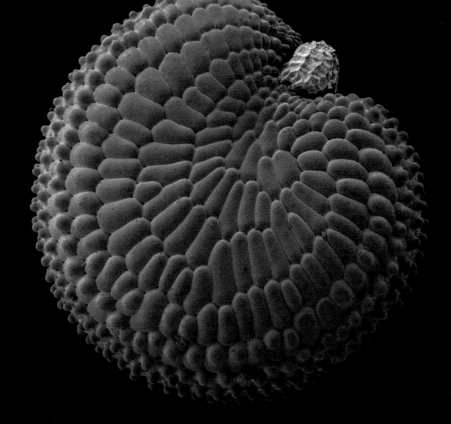

Calandrinia eremaea (Portulacaceae) – twining purslane; collected in South Australia – the seed is shed from capsules and has no obvious adaptations to a specific mode of dispersal. The conspicuous swelling at the hilum (yellow) might function as an elaiosome to attract ants but its texture and natural colour (black) suggest that it is probably a rudimentary aril that has lost its function; seeds 0.56mm in diameter

Euphorbia sp. (Euphorbiaceae) – spurge;
collected in Lebanon – seed, 3mm long

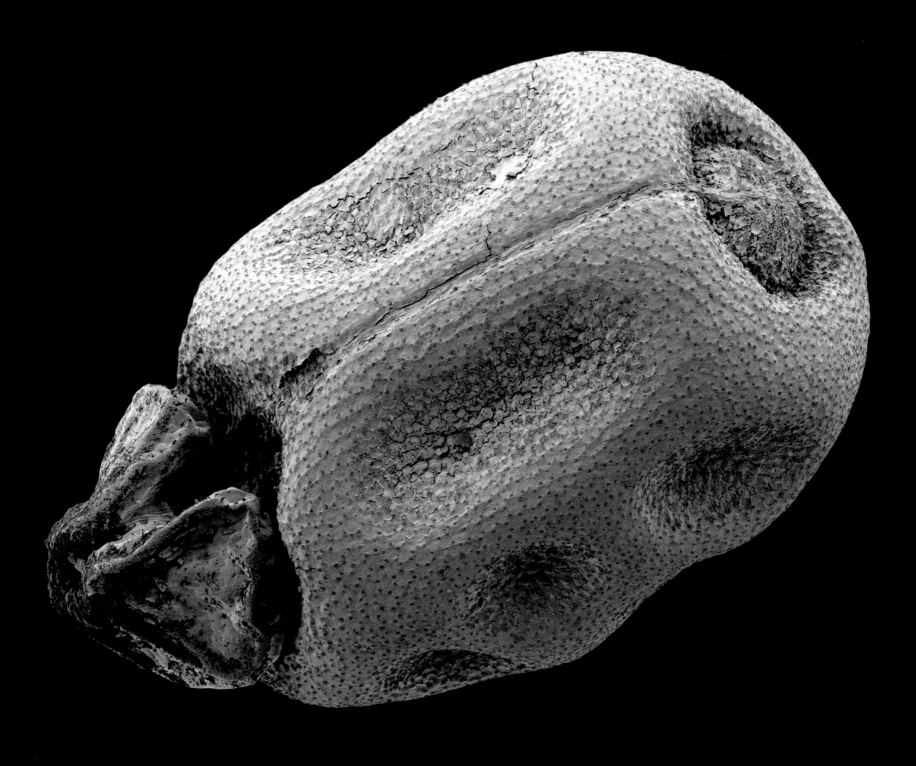

Euphorbia peplus (Euphorbiaceae) – petty spurge; collected in the UK; 1.6mm long – *Euphorbia* seeds commonly bear an aril (elaiosome) to attract ants. In the entire family the arils are formed by a swelling of the outer integument around the micropyle (exostome); the central micropylar canal persists and is often still visible in the mature seed

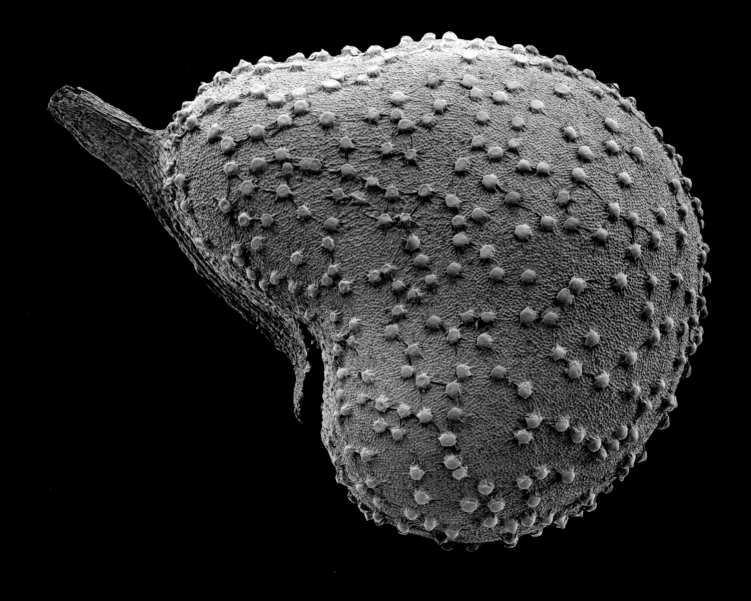

Abutilon angulatum (Malvaceae) – collected in Botswana – seed with no obvious adaptations to a specific mode of dispersal; part of the funiculus is still attached; 2.4mm long

opposite: *Abutilon pictum* cultivar (Malvaceae) – flower

Agrostemma githago (Caryophyllaceae) – corncockle; collected in the UK – flower and seed; the seed has no obvious adaptations to a specific mode of dispersal but displays the intricate surface pattern typical of the pink family: the papillose cells of the seed coat are interlocked like the pieces of a jigsaw puzzle. Dispersal is achieved when the capsules expel the seeds as they sway in the wind on their flexible stalks; seed 3.7mm in diameter

Alcea pallida (Malvaceae) – pale hollyhock; native to central and south-eastern Europe – individual fruitlet 5.3mm in diameter

opposite: mature fruit; the (schizocarpic) fruit develops from a syncarpous ovary and splits at maturity into one-seeded segments (fruitlets), each corresponding to a single carpel

Argemone platyceras (Papaveraceae) – prickly poppy; native to the USA and Mexico – flower and seed; no obvious adaptations to a specific mode of dispersal, which occurs when the capsules expel the seeds in the wind; 2.1mm in diameter

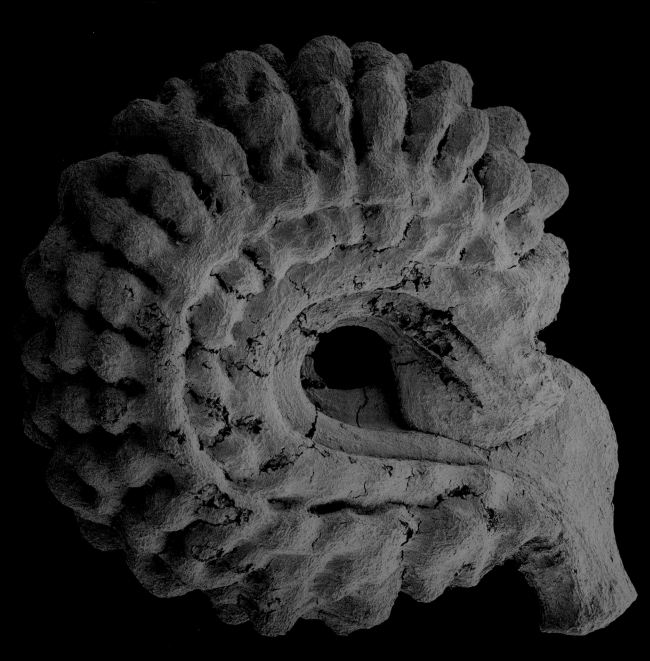

Calendula officinalis (Asteraceae) – pot marigold; popular garden plant, origin unclear – flower and fruit (achene); the curved fruits of this species vary tremendously in size and shape. Three different types of fruits are present in the same seed head, a phenomenon called heterocarpy (Greek: *heteros* = different + *karpos* = fruit). The large, spiny outermost ones are adapted to both animal and wind dispersal; those in the middle are spoon-shaped and adapted to wind and water dispersal; the innermost fruits (illustrated) resemble worms or insect larvae, possibly to trick birds into picking them up; the fruit illustrated is 3.9mm in diameter

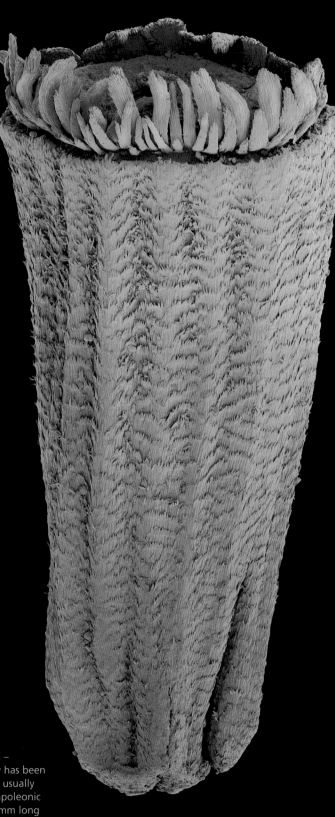

Cichorium intybus (Asteraceae) – chicory; collected in the UK – flower and fruit (achene with rudimentary pappus) – chicory has been cultivated since ancient times for its edible leaves, which are usually blanched to reduce their bitterness. In Europe during the Napoleonic blockade, chicory roots were used as a coffee substitute; 2.4mm long

Lophophora williamsii (Cactaceae) – peyote; native to the Chihuahuan Desert of Texas and northern Mexico – like most berry-borne seeds those of the peyote show no obvious adaptations to a specific mode of dispersal; 1.3mm long

opposite: apex of plant with flower and fruits (berries). Peyote has long been a sacred plant to native Americans, who use it ceremonially. The small spineless cactus contains more than 50 alkaloids, the most important being mescaline, which has psychotomimetic effects similar to those of LSD

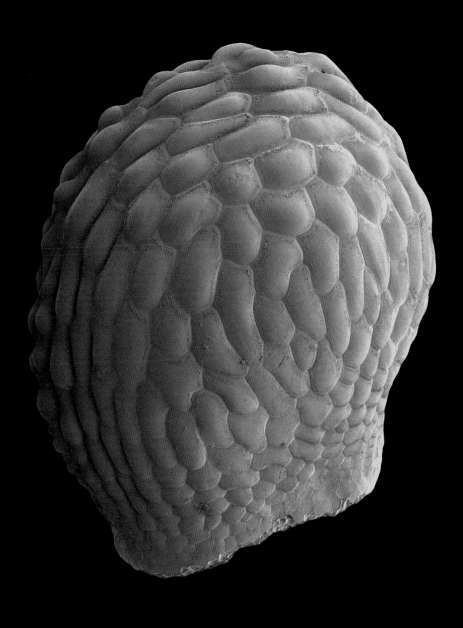

Melocactus neryi (Cactaceae) – turk's-cap cactus; native to South America – borne in berries the seeds show no obvious adaptations to a specific form of dispersal; 1.3mm in diameter

opposite: apex of plant with fruits. Melocacti are among the most fascinating cacti. While most cacti do not change their appearance as they mature, melocacti (and some other genera) produce a specialised flower-bearing *cephalium* at the top when they reach flowering age. As soon as the cephalium is formed the vegetative growth of the plant body stops and only the cephalium continues to grow and produce flowers

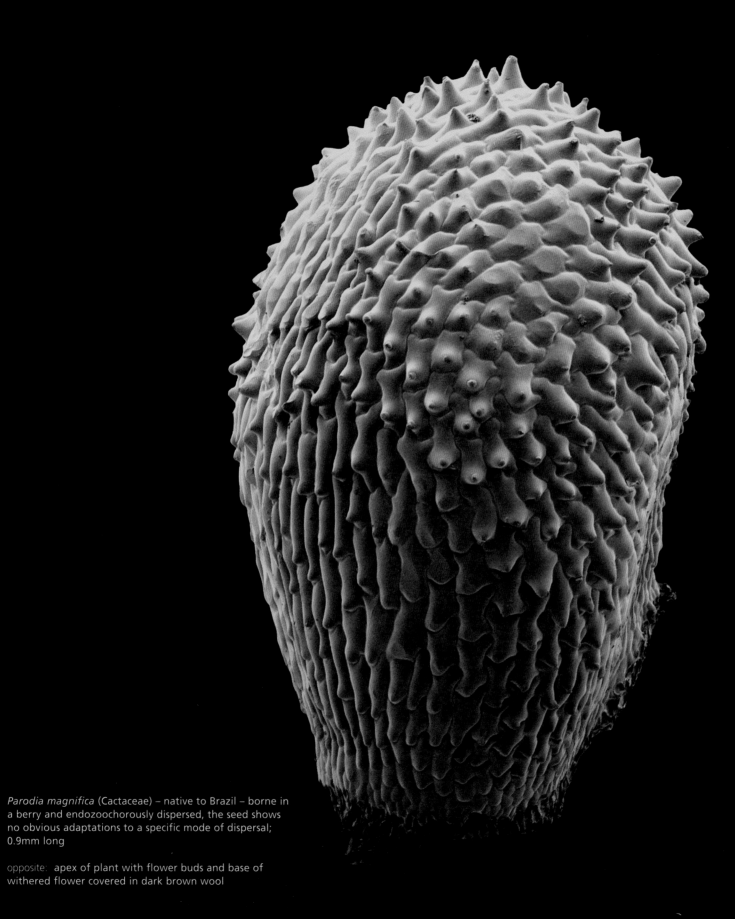

Parodia magnifica (Cactaceae) – native to Brazil – borne in a berry and endozoochorously dispersed, the seed shows no obvious adaptations to a specific mode of dispersal; 0.9mm long

opposite: apex of plant with flower buds and base of withered flower covered in dark brown wool

Lychnis flos-cuculi (Caryophyllaceae) – ragged robin; collected in the UK – flower and seed with no obvious adaptations to a specific mode of dispersal but displaying the intricate surface pattern typical of the pink family; dispersal occurs when the capsules expel the seeds as they sway in the wind on their flexible stalks; 0.9mm long

Echium vulgare (Boraginaceae) – viper's bugloss; collected in the UK – single-seeded nutlet; like many members of the borage family, the ovary of this species is deeply four-lobed and splits at maturity into four single-seeded nutlets, each 2.7mm long

opposite: close-up of flower

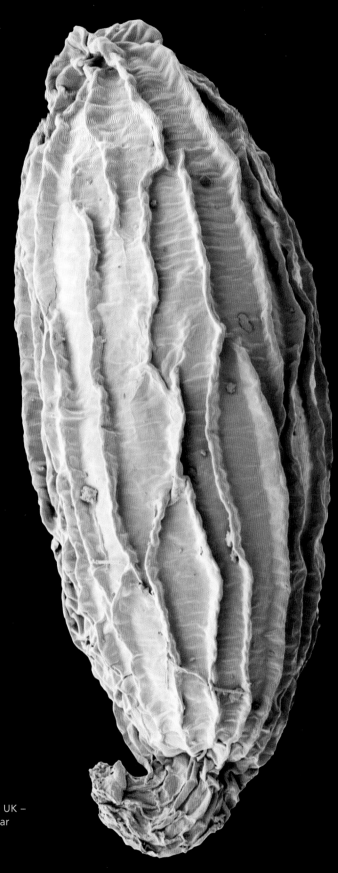

Sedum album (Crassulaceae) – white stonecrop; collected in the UK – wind-dispersed dust seed with typical spindle-shape and irregular longitudinal ridges to increase air resistance; 0.6mm long

opposite: inflorescence

Silene gallica (Caryophyllaceae) – small-flowered catchfly; collected in the UK – seed with no obvious adaptations to a specific mode of dispersal but displaying the intricate surface pattern typical of the pink family: the convex cells of the seed coat are interlocked like the pieces of a jigsaw puzzle. Dispersal occurs when the capsules expel the seeds as they sway in the wind; 1.5mm long

opposite: close-up of flower

Silene maritima (Caryophyllaceae) – sea campion; collected in the UK – flower and seeds; the seeds show no obvious adaptations to a specific mode of dispersal but display the intricate surface pattern typical of the pink family; dispersal is achieved when the capsules expel the seeds as they sway in the wind; seed 1.3mm long

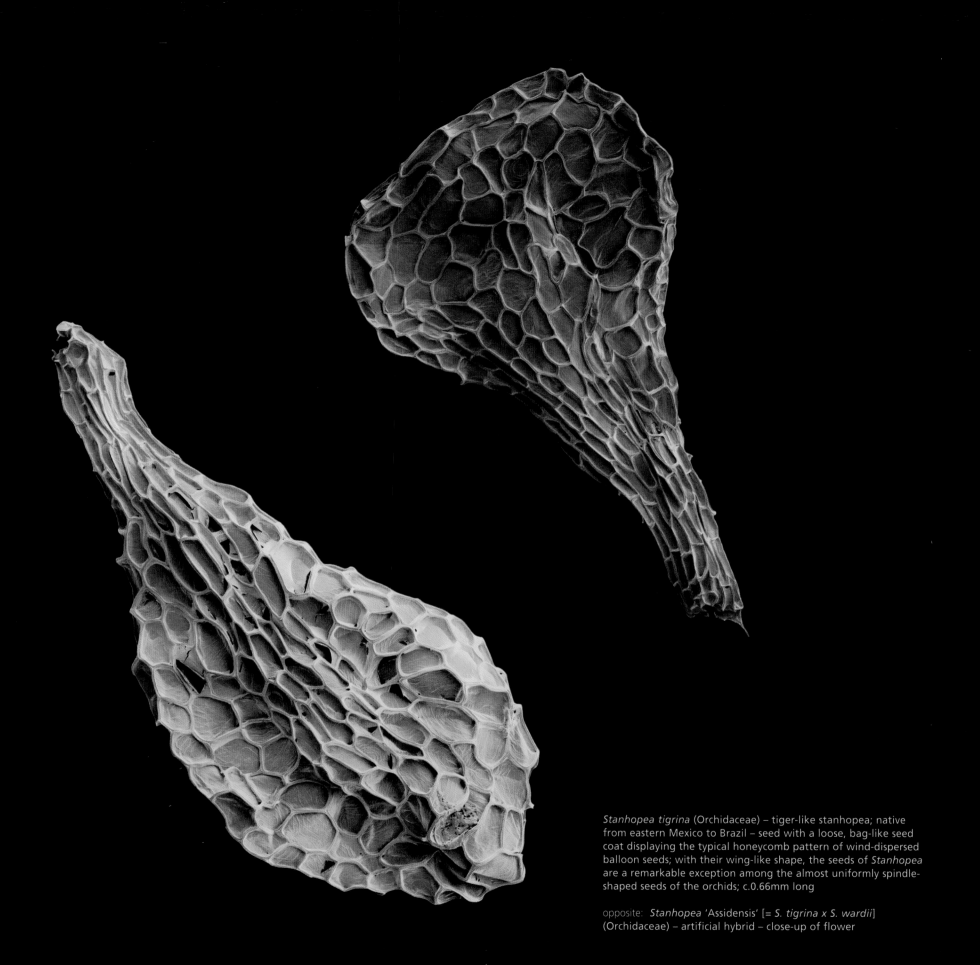

Stanhopea tigrina (Orchidaceae) – tiger-like stanhopea; native from eastern Mexico to Brazil – seed with a loose, bag-like seed coat displaying the typical honeycomb pattern of wind-dispersed balloon seeds; with their wing-like shape, the seeds of *Stanhopea* are a remarkable exception among the almost uniformly spindle-shaped seeds of the orchids; c.0.66mm long

opposite: *Stanhopea* 'Assidensis' [= *S. tigrina* x *S. wardii*] (Orchidaceae) – artificial hybrid – close-up of flower

Seeds of sundews (Droseraceae) – left: *Drosera capillaris* – pink sundew; native to the eastern United States – seed 0.6mm long; right: *Drosera natalensis* – Natal sundew; native to South Africa, Mozambique and (allegedly) Madagascar – seed 0.8mm long; the seeds of both species display the typical spindle-shape and reticulate pattern of wind-dispersed dust seeds

opposite: *Drosera capillaris* (Droseraceae) – pink sundew; native to the eastern United States – inflorescence

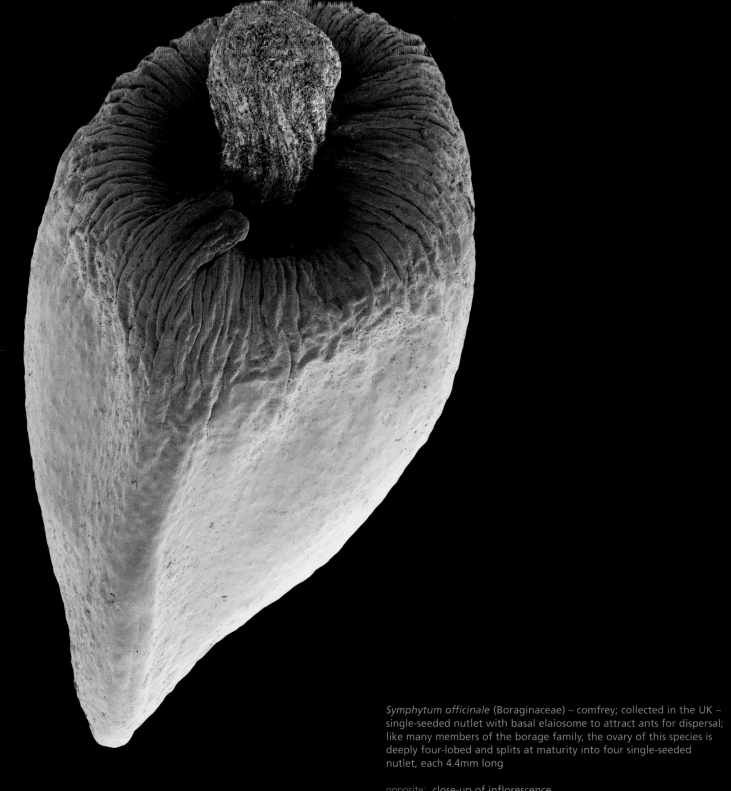

Symphytum officinale (Boraginaceae) – comfrey; collected in the UK – single-seeded nutlet with basal elaiosome to attract ants for dispersal; like many members of the borage family, the ovary of this species is deeply four-lobed and splits at maturity into four single-seeded nutlet, each 4.4mm long

opposite: close-up of inflorescence

Thlaspi arvense (Brassicaceae) – field penny-cress; collected in the UK – seed; no obvious adaptations to a specific mode of dispersal other than its flattened shape, which may assist wind dispersal; 2.1mm long

opposite: fruit; when hit by a drop of water or shaken by strong wind the flexible stalk bends over and bounces back, ejecting the two papery sides of the fruit with the seeds inside

Anarrhinum orientale (Plantaginaceae) – native to the Mediterranean – flower and seed; seed displaying the typical honeycomb pattern of wind-dispersed balloon seeds. As in the similar seeds of the closely-related foxglove (*Digitalis* spp.), the outer tangential wall of the cells of the seed coat has totally disappeared at maturity, exposing the radial walls that form the honeycomb; seed 1.25mm long

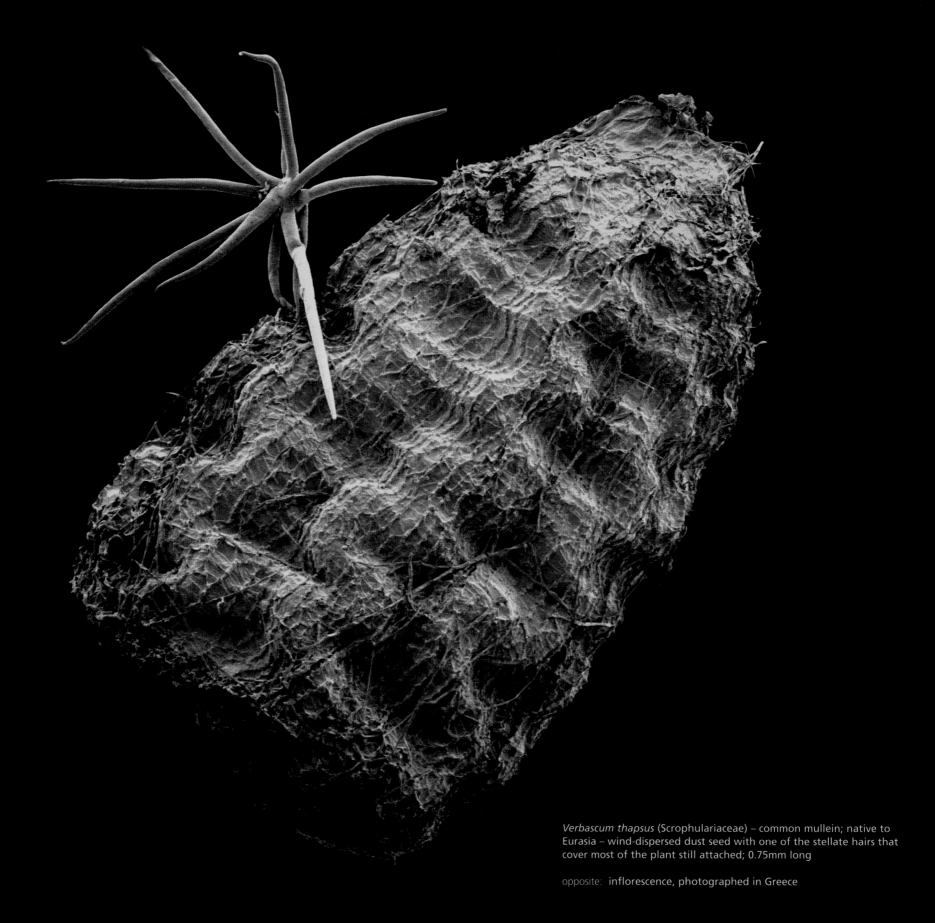

Verbascum thapsus (Scrophulariaceae) – common mullein; native to Eurasia – wind-dispersed dust seed with one of the stellate hairs that cover most of the plant still attached; 0.75mm long

opposite: inflorescence, photographed in Greece

Drosophyllum lusitanicum (Droseraceae) – Portuguese sundew; native to Spain and Portugal – seed; 2.7mm long

opposite: tip of a leaf with sticky glandular hairs poised to catch insects

Verbascum sinuatum (Scrophulariaceae) – wavyleaf mullein; native to the Mediterranean – wind-dispersed dust seed; 1.6mm long

opposite: close-up of flower

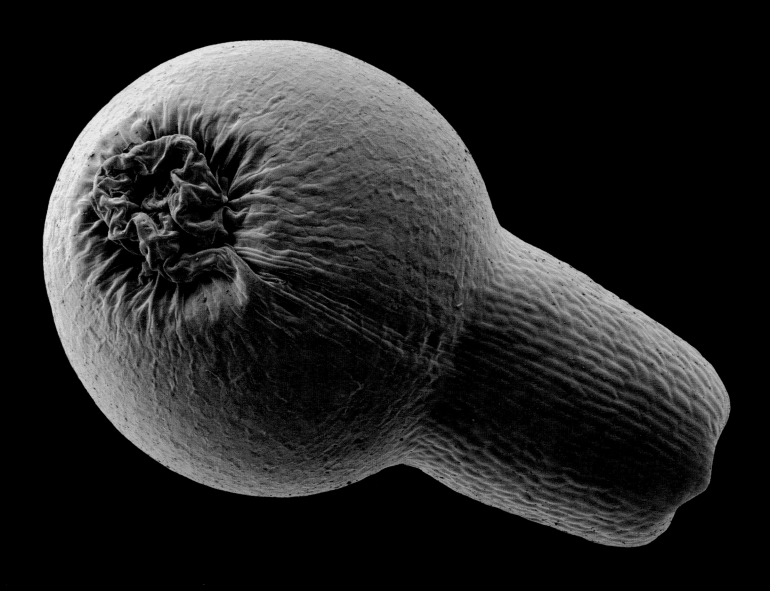

Lachenalia lutea (Hyacinthaceae) – yellow Cape cowslip; collected in South Africa – seed; an unusual shape created by the hollow and (under a light microscope) semi-transparent, tubular extension on the right, the function of which is unclear; it may assist wind dispersal or act as a buoyancy device during heavy rainfall, or act as an elaiosome to attract ants for dispersal; seed 1.1mm long

opposite: *Lachenalia peersii* (Hyacinthaceae) – bekkies (Afrikaans); native to the Cape Province, South Africa – close-up of flowers photographed in the Fernkloof Nature Reserve, Hermanus

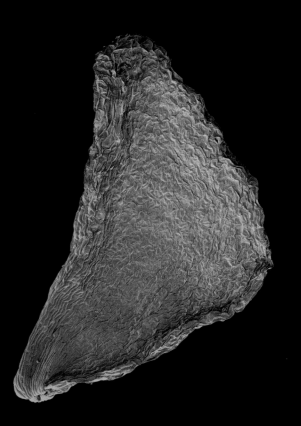

Fritillaria meleagris (Liliaceae) – snake's head fritillary; collected in the UK – flower and seed

opposite: loculicidal capsule and seed; the extremely flattened shape of the seed is an adaptation to facilitate wind dispersal; 3.7mm long

opposite: *Allium ampeloprasum* (Alliaceae) – wild leek; native to Eurasia and North Africa – seeds; their flattened shape indicates wind dispersal; 2.9mm long

above: cluster of mature capsules

Seeds: Time Capsules of Life

Silene dioica (Caryophyllaceae) – red campion; collected in the UK – capsules and seeds the latter have no obvious adaptations to a specific mode of dispersal but display the intricate surface pattern typical of the pink family; dispersal occurs when the capsules expel the seeds as their flexible stalks sway in the wind; seed 1.2mm long

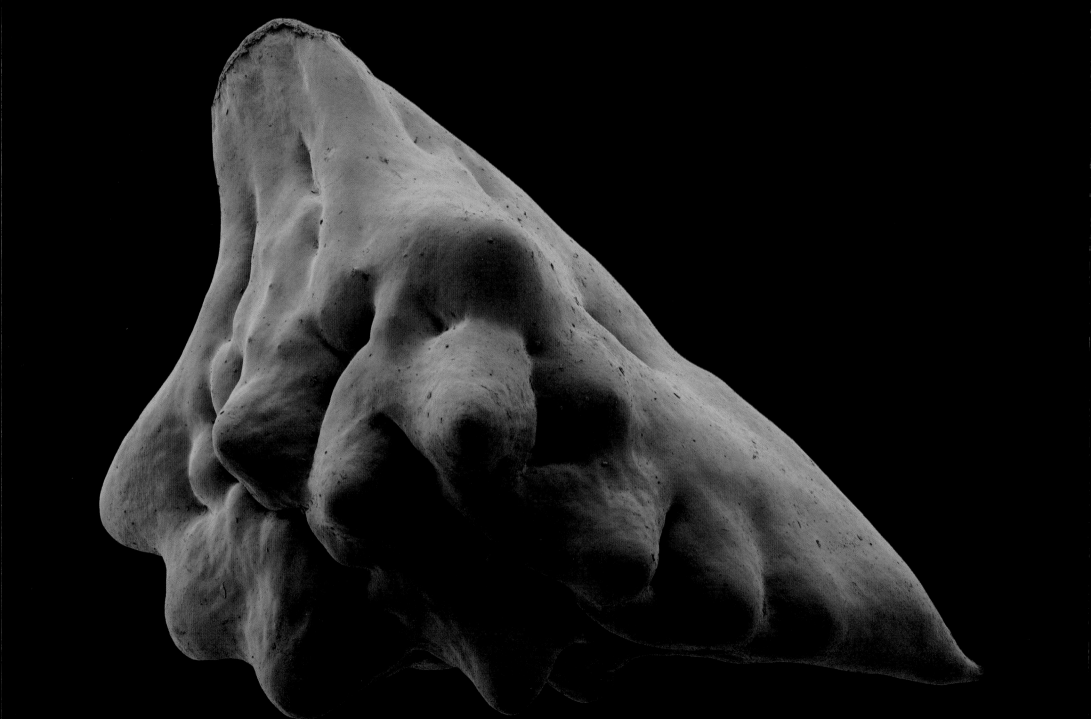

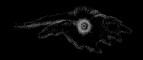

APPENDICES

Lappula spinocarpos (Boraginaceae) – single-seeded nutlet; like many members of the borage family, the ovary of this species is deeply four-lobed and splits at maturity into four single-seeded nutlets

GLOSSARY

EXPLANATORY NOTES

Explaining the scientific background of seeds requires a certain number of technical terms. When used for the first time in the text, technical terms are italicised. To avoid spoiling the flow of the text a glossary is included for those readers not yet familiar with these terms. The common names of plants are used whenever possible, but their Latin names are indispensable because they are unique and understood by naturalists worldwide, whatever their mother language. Common names are different in every language and some species may have several common names, and one common name may refer to several different species. Latin names generally consist of two parts, the genus name and the species name (e.g. *Liriodendron tulipifera*). A full Latin name includes its formal describer(s) at the end as in the *Index of Plants Illustrated* that follows. A group of closely-related species forms a genus and a group of closely-related genera forms a family (e.g. Magnoliaceae). With the application of molecular techniques to elucidate the natural relationships of plants, their classification – especially in the case of the flowering plants – has undergone profound changes. The boundaries of many long-established families have changed and some have been split (most recently, for example, the Scrophulariaceae). For the familial classification we have adopted Peter Stevens's system (Stevens, P.F. (2001 onwards). Angiosperm Phylogeny Website. Version 6, May 2005), which largely follows the latest classification of the Angiosperm Phylogeny Group, an international team of scientists jointly researching the natural relationships of the angiosperms. The website of IPNI (International Plant Names Index; www.ipni.org) and the W3TROPICOS website of the Missouri Botanical Garden (http://mobot.mobot.org/W3T/Search/vast.html) proved very helpful in verifying Latin names.

With the exception of a few selected images and the diagrams drawn by Elly Vaes, the majority of photographs included here are the original work of the authors. Photographs were taken with Nikon digital cameras (model D100 and D70) using 60mm micro nikkor and 35-105 macro nikkor lenses. Digital scanning electron micrographs were produced with a Hitachi S-4700 Scanning Electron Microscope (SEM). Rob Kesseler subsequently coloured the original black-and-white SEM-images but left them unaltered in any other way. The choice of colours was inspired by the natural colours of the plant or its flowers, the structure and function of the seed coat, or simply by the intuition of the artist. Seed images were primarily selected for their aesthetic qualities. The broadest possible variety of seeds were chosen to reveal the magnificent diversity of structures and functional adaptations.

The seeds illustrated in this book have mainly been sourced from the collections of the Millennium Seed Bank but also Kew's Herbarium Seed Collection (now housed at the Millennium Seed Bank), the drift seed collection of Kew's Centre for Economic Botany, and the carpological collection of the Kew Herbarium.

Frequently used abbreviations: sp. = species (singular); spp. = species (plural)

achene (Greek: *a* + *khainein* = to yawn): a small indehiscent, usually single-seeded fruit with a dry pericarp that is contiguous to the seed but distinguishable from the seed coat, e.g. sunflower (*Helianthus annuus*, Asteraceae).

aerial seed bank: see serotiny.

anatropous: see anatropy.

anatropy (Greek: *ana-* = upon, upwards + *tropos* = direction, turn): condition in which an ovule with a straight longitudinal axis is inverted by 180° so that the body of the ovule is parallel to the funiculus. As a result of this inversion, the point where the funiculus is attached (hilum) moves to lie next to the morphological apex (micropyle). The anatropous ovule is the most common type of ovule in the angiosperms and more than 80% of all families are exclusively anatropous.

anemoballism (Greek: *anemos* = wind + *ballistes*, from *ballein* = to throw): form of dispersal in which the diaspores are subject to indirect effects of wind, i.e. the wind does not transport the diaspore directly but exerts its influence on the fruit. The fruit (mostly a capsule) is usually exposed on a long flexible stalk that throws out the diaspores as it sways in the wind, e.g. sacred lotus (*Nelumbo nucifera*, Nelumbonaceae), poppy (*Papaver rhoeas*, Papaveraceae).

anemoballist: a plant dispersing its diaspores by anemoballism. See also anemoballism.

anemochory (Greek: *anemos* = wind + *chorein* = to disperse, to wander): dispersal of fruits and seeds by wind.

angiosperms (Greek: *angeion* = vessel, small container + *sperma* = seed): division of the seed plants (spermatophytes) that bear ovules and seeds within closed megasporophylls (carpels); in contrast to gymnosperms that have exposed ovules and seeds, borne "naked" on the megasporophylls. Angiosperms are distinguished by a unique process of sexual reproduction called "double fertilization". According to the number of leaves (cotyledons) present in the embryo two major groups are distinguished, the Monocotyledons and the Dicotyledons. Angiosperms are commonly referred to as "flowering plants" even though the reproductive organs of some gymnosperms are also borne in structures that fulfil the definition of a flower.

anisocotyly (Greek: *anisos* = unequal + *cotyledon* = seed leaf): term referring to the difference in size between the cotyledons (seed leaves) of a dicotyledonous embryo.

anther (Medieval Latin: *anthera* = pollen, derived from Greek: *antheros* = flowery, from *anthos* = flower): the pollen-bearing part of a microsporophyll (stamen) of the angiosperm. An anther consists of two fertile halves called *thecae*, each bearing two pollen sacs (= microsporangia) which usually dehisce with longitudinal slits, or pores (e.g. Ericaceae) or by valves (e.g. Lauraceae). The two thecae are connected by a sterile part called the *connective*, which is also the point where the anther is fixed to the filament.

antheridium (Latin: small anther; *anther* referring to the pollen-bearing plant of the angiosperms): male sexual organ of a male or bisexual gametophyte producing and containing the male gamete(s). Antheridia are fully developed in mosses, ferns and fern allies in the broadest sense, but absent in gymnosperms and angiosperms.

anthocarp (Greek: *anthos* = flower + *karpos* = fruit): a fruit in which not only the gynoecium but also other floral parts have undergone a marked development during post-fertilisation to aid in the dissemination of the seed.

anthocarpous fruit = anthocarp.

anthophytes (Greek: *anthos* = flower + *phyton* = plant): literally *flowering plants*, a term often used as a synonym for angiosperms. However, anthophytes also include some gymnosperms, the extinct cycad-like Bennettitales, the closely related *Pentoxylon* and the present-day Gnetales order (comprising the three genera *Ephedra*, *Gnetum*, and *Welwitschia*).

anthropochory (Greek: *anthropos* = human being + *chorein* = to disperse, to wander): dispersal of fruits and seeds by man.

antipodal cells (Greek: *anti-* = anti- + *podos* = foot): three cells in the embryo sac (megagametophyte) of the angiosperms. The Greek meaning "with the feet opposite" refers to the position of the antipodal cells diametrically opposite the egg apparatus (= egg cell plus two synergids).

apocarpous gynoecium (Greek: *apo* = being apart from + *karpos* = fruit): gynoecium (the female parts of a flower) consisting of two or more separate carpels, each carpel forming an individual pistil.

apomixis (Greek: *apo* = away from, without + *mixis* = intercourse, mingling): a form of asexual reproduction whereby the ovule develops into a seed without meiosis or fertilization. In a wider sense the term apomixis is sometimes also deemed to include any type of asexual reproduction including vegetative propagation by suckers, offsets etc.

archegonium (Modern Latin, from Greek: *arkhegonos* = offspring; from *arkhein* = to begin + *gonos* = seed, procreation): often flask-shaped, multi-cellular female sexual organ of a female or bisexual gametophyte producing and containing the female egg cell(s). Archegonia are fully developed in mosses, ferns and fern allies in the broadest sense, but only rudimentary in gymnosperms. In angiosperms true archegonia are absent (with the three-celled egg apparatus as the homologue).

archespore (Greek: *arkhein* = to begin + *sporos* = germ, spore): in mosses, ferns and fern allies, gymnosperms and angiosperms the sporogenous cell or tissue inside a sporangium that divides mitotically to produce the spore mother cells. In angiosperms, for example, the archespore of the pollen sac gives rise to the pollen mother cells which, following meiosis, each produce four pollen.

aril (Latin: *arillus* = grape seed): edible seed appendages of various origins in gymnosperms and angiosperms. Arils usually developed as a reward for animal dispersers.

atropous: see atropy.

atropy (Greek: *a-* = negating prefix + *tropos* = direction, turn): condition of an ovule in which funiculus, chalaza and micropyle lie in a straight line that follows the longitudinal axis of the ovule. As a result, atropous ovules lack a raphe and the hilum is situated adjacent to the chalaza. Atropous ovules are typical of gymnosperms which is the reason why they were often thought to be the primitive condition in angiosperms. However, the atropous ovules of angiosperms seem to be derived from anatropous ovules and are found in a diverse range of about 20 angiosperm families, e.g. Araceae, Commelinaceae, Haemodoraceae, Juglandaceae, Piperaceae, Polygonaceae, Proteaceae, Urticaceae. Another term which is often used for atropy is orthotropy. However, the latter means a *straight turn* which is a contradiction in terms.

autochory (Greek: *autos* = self + *chorein* = to disperse, to wander): self-dispersal.

ballistic dispersal: mode of dispersal by which the diaspores are actively or passively catapulted away from the plant. This can either happen arbitrarily or be set off by an external trigger.

bitegmic ovule: an ovule covered by two separate envelopes (integuments).

bract: a reduced or rudimentary leaf in the region of the flower or inflorescence. Bracts can be either small, green and inconspicuous or rather large and conspicuously coloured.

calyx (Greek: *kalyx* = cup): all the sepals of a flower, i.e. the outer whorl of floral leaves in a perianth.

campylotropous seed: see campylotropy

campylotropy (Greek: *kampylos* = curved + *tropos* = direction, turn): condition in which ovules and seeds have a curved longitudinal axis, usually resulting in a curved embryo. Embryos of campylotropous seeds can be teic as long as the seed, resulting, in a larger, more successful seedling.

capsule: a dehiscent fruit developing from a syncarpous gynoecium (i.e. composed of more than one carpel) and dispersing the seeds by opening the pericarp.

Carboniferous: geological time period 354-290 million years ago.

carpel (Modern Latin: *carpellum* = little fruit; originally from Greek: *karpos* = fruit): in angiosperms a fertile leaf that encloses one or more ovules (megasporophyll). Carpels are usually differentiated into an ovule-bearing part (ovary), a style and a stigma. The carpels of a flower can be either separate from each other to form an apocarpous gynoecium or united to form a syncarpous gynoecium.

caryopsis (Greek: *karyon* = walnut or any nut, kernel + *-opsis* = resemblance): the traditional name of the fruit (nut) of the members of the grass family (Poaceae). The caryopsis is very similar to the achene. The only difference is that in a caryopsis the pericarp is not distinguishable from the seed coat except under high magnification.

central cell: the large binucleate cell in the middle of the mature embryo sac (megagametophyte) of the angiosperms, containing the two polar nuclei.

chalaza (Greek: *khalaza* = hard lump, hailstone): a topographic term that describes the region at the base of the ovule where the integument(s) and the nucellus are joined. Funiculus, chalaza and raphe form a continuous tissue without any sharp delimitations.

clubmoss: common name for a member of the Lycophyta, a group of seedless, spore-producing vascular plants. In the Carboniferous time period (354-290 million years ago) tree-like clubmoss relatives such a *Lepidodendron* and *Sigillaria* together with horsetail relatives were the major components of the giant forests that thrived in the extensive swamps occupying large parts of our planet. Today, Lycophytes are represented by about 1,280 species of herbaceous plants such as clubmosses (*Lycopodium* spp.), selaginellas (*Selaginella* spp.) and the aquatic quillworts (*Isoetes* spp.).

coma (Latin = hair, from Greek: *come* = hair): a tuft of hairs in plumed seeds that facilitates wind dispersal.

compound fruit: a fruit derived from more than one flower. Most modern textbooks apply this meaning to multiple fruit, a term which should, however, correctly be used for fruits developing from flowers with an apocarpous gynoecium; see also explanation under aggregate fruit.

conifers (Latin: *conus* = cone + *ferre* = to carry, to bear): group of the gymnosperms generally distinguished by needle- or scale-like leaves and unisexual flowers borne in cones. Well-known examples of conifers are pines, spruces and firs.

corolla (Latin *corolla* = small garland or crown): all the petals of a flower, i.e. the inner whorl of floral leaves in a perianth.

cotyledon (Greek: *kotyle* = hollow object; alluding to the often spoon- or bowl-shape of the seed leaves): the first leaf (in Monocotyledons) or pair of leaves (in Dicotyledons) of the embryo.

Cretaceous: geological time period 65-142 million years ago.

cryptogams (Greek: *kryptos* = hidden + *gamein* = to marry, to copulate): old collective term referring to all plants without recognizable flowers. Cryptogams include algae, fungi (although they are not really plants), mosses, ferns and fern allies. The Greek meaning "those who copulate in secrecy" refers to the absence of flowers as obvious indicators of sexual propagation.

cupule (Latin: *cupula* = little cask): loose cup-like structure surrounding the ovules of the earliest seed plants (pteridosperms).

cycads (Greek: *kykas* = palm, alluding to their palm-like look): ancient gymnosperms superficially resembling palms. Cycads are woody plants generally distinguished by thick, unbranched trunks, large, palm-like pinnate leaves and big cones. Living fossils, cycads were an important source of food for the dinosaurs.

cypsela (Greek: *kypselé* = box, hollow vessel): a single-seeded fruit with longitudinally oriented awns, bristles, feathers or similar structures derived from accessory parts of the flower or inflorescence. Cypselas are typically found in the Asteraceae and Dipsacaceae but also in some Proteaceae and other families.

Devonian: geological time period 417-354 million years ago.

diaspore (Greek: *diaspora* = dispersion, dissemination): the smallest unit of seed dispersal in plants. Diaspores can be seeds, fruitlets of compound or schizocarpic fruits, entire fruits or seedlings (e.g. in mangroves).

Dicots = Dicotyledons.

Dicotyledons (Greek: *di* = two + cotyledon): one of the two major groups of the angiosperms distinguished by the presence of two leaves (cotyledons) in the embryo. Other typical characters of the Dicotyledons are reticulate leaf venation, floral organs usually in fours or fives, vascular bundles arranged in a circle, a persistent primary root system developing from the radicle, and secondary thickening (present in trees and shrubs, usually absent in herbaceous plants). The Dicotyledons were long considered a homogeneous entity. Recently they have been split into two groups (magnoliids and Rosidae (eudicots).

dioecious: see dioecy

dioecy (Greek: *di* = double + *oikos* = house): (1) the formation of male and female sexual organs on separate gametophytes (e.g. in some mosses and ferns), (2) in seed plants the formation of male and female flowers on separate individuals.

dormancy (Latin: *dormire* = to sleep): generally referring to a quiescent period in the life of a plant. When referring to seeds, dormancy summarizes the various mechanisms which ensure that seeds do not germinate immediately, even under the most favourable conditions.

egg apparatus: in angiosperms the egg cell plus the two synergids at the micropylar end of the embryo sac (megagametophyte).

elaiosome (Greek: *elaion* = oil + *soma* = body). literally meaning "oil body"; a general ecological term referring to fleshy and edible appendages of diaspores, usually in the context of ant dispersal.

embryo (Latin: *embryo* = unborn foetus, germ, originally from Greek: *embryon*: *en-* = in + *bryein* = to be full to bursting): in plants the young sporophyte developing from the egg cell after fertilization.

embryo sac: the female gametophyte (megagametophyte) of the angiosperms. The embryo sac develops from the functional megaspore normally through three mitotic divisions, producing a total of eight nuclei distributed over seven cells: an egg apparatus composed of the egg cell and two synergids, three antipodal cells and one binucleate central cell.

endocarp (Greek: *endon* = inside + *karpos* = fruit): the innermost layer of the fruit wall (pericarp) forming the hard stone in drupes.

endosperm (Greek: *endon* = inside + *sperma* = seed): nutritive tissue in seeds. Generally, the term endosperm refers only to the nutritive tissue in the seeds of the angiosperms where it forms a (usually) triploid tissue as a result of the double fertilization. The food storage of gymnosperm seeds consists of the haploid tissue of the megagametophyte. To distinguish the two different types, the nutritive tissues of gymnosperms and angiosperms are called "primary endosperm" and "secondary endosperm" respectively.

endostome (Greek: *endon* = inside + *stoma* = mouth, the "inner mouth"): in a bitegmic angiosperm ovule the micropylar opening of the inner integument.

endozoochory (Greek: *endon* = inside + *zoon* = animal + *chorein* = to disperse, to wander): dispersal of the diaspores of a plant by being eaten and carried inside the gut of animals (and humans); the usually hard seeds or endocarps pass through the intestines undamaged and are deposited with the faeces.

epicarp (Greek: *epi* = on, upon + *karpos* = fruit): the outermost layer of the fruit wall (pericarp), mostly a soft skin or leathery peel.

epizoochory (Greek: *epi* = on, upon + *zoon* = animal + *chorein* = to disperse, to wander): dispersal of diaspores on the surface of the body of an animal. Epizoochorous diaspores adhere to the fleece, coat or feathers of animals or birds or the clothes of humans by hooks or sticky substances.

exostome (Greek: *exo* = outside + *stoma* = mouth, the "outer mouth"): in a bitegmic angiosperm ovule the micropylar opening of the outer integument.

family: one of the main units in the hierarchical system of the taxonomic classification of living organisms. The major classification units are (in descending order) class, order, family, genus and species.

filament (Latin: *filum* = thread, string): the stalk of a stamen.

flagellum (plural *flagellae*; Latin: *flagellum* = little whip): thread-like projections which enable cells to move.

floret (Latin diminutive of *flor* = flower): part of the inflorescence in grasses (spikelet) formed by the lower and upper bract (lemma and palea) typically enclosing a single flower. The term floret is sometimes also applied to the flowers of the Cyperaceae and Asteraceae.

flower: a reproductive short shoot with determinate growth bearing at least one of the sexual reproductive organs (male or female). This definition of a flower applies to the reproductive structures of both angiosperms (flowering plants) and gymnosperms.

flowering plants: meaning differs regionally depending on the local definition of a flower. In continental Europe considered to comprise both gymnosperms and angiosperms, in Anglo-America and UK applied only to angiosperms. In a strict scientific sense, "flowering plants" are circumscribed as defined under "anthophyta".

follicle (Latin: *folliculus* = little bag, diminutive of *follis* = bellows): a fruit or fruitlet derived from a single carpel dehiscing along one (usually the ventral) suture, e.g. the fruitlets of the marsh marigold (*Caltha palustris*, Ranunculaceae).

frugivore (Latin: *frug-* = fruit + *vorare* = to swallow, devour): fruit-eating animals.

fruitlet: a separate dispersal unit of a fruit that may be (1) a carpel or half-carpel of a mature schizocarpic fruit, (2) a single carpel of a mature multiple fruit, or (3) a mature multicarpellate ovary of a compound fruit.

functional megaspore: of the four megaspores resulting from the meiosis of the megaspore mother cell only one, the functional megaspore, survives to give rise to the embryo sac; the remaining three megaspores are aborted.

funiculus/funicle (Latin: *funiculus* = slender rope): the stalk by which an ovule or seed is connected to the placenta in the ovary. The funiculus acts like an umbilical cord, supplying the developing ovule and seed with water and nutrients from the parent plant.

gamete (Greek: *gametes* = spouse): haploid male or female germ cells. Male and female gametes fuse on copulation. In contrast to spores, they can only give rise to a new individual or generation after they have fused with a gamete of the opposite sex.

gametophyte (Greek: *gametes* = spouse + *phyton* = plant): the generally haploid generation in a plant's life cycle producing gametes. Examples are the prothallus of the ferns or the megaprothallium (= female gametophyte) and germinated pollen grain (= microprothallium = male gametophyte) of the seed plants.

geocarpy (Greek: *ge* = earth + *karpos* = fruit): a condition in plants whose fruits ripen underground, e.g. peanut (*Arachis hypogaea*, Fabaceae).

germplasm: general term referring to cells or sexual and asexual reproductive structures (e.g. seeds) from which new plants can be regenerated.

Gnetales: heterogeneous group of gymnosperms comprising just three families with three genera (*Gnetum, Ephedra, Welwitschia*) and a total of 95 species.

gymnosperms (Greek: *gymnos* = naked + *sperma* = seed): inhomogeneous group of seed plants bearing their ovules on open megasporophylls (or ovuliferous scales in conifers) and not in closed megasporophylls (= carpels) as the case in angiosperms. Gymnosperms comprise three distantly related groups: conifers (8 families, 69 genera, 630 species), cycads (3 families, 11 genera, 292 species) and Gnetales (3 families, 3 genera, 95 species).

gynoecium (Greek: *gyne* = woman + *oikos* = house): all the carpels in a flower, irrespective of whether they are united or separate.

gynophore (Greek: *gyne* = female, woman + *phorein* = to carry): a stalk-like elongation of the flower axis elevating the gynoecium above the point where sepals, petals and anthers insert.

haustorium (Modern Latin, from classical Latin: *haustus* = absorption): general term referring to absorbing organs that function as suckers to provide the plant or fungus or a specific organ with water and/or nutrients (e.g. an outgrowth of stem, root, hyphae etc.), mostly applied to the specialized structure with which parasites absorb water and nutrients from their host plants. Haustoria can also be formed by the suspensor, the synergids, the antipodal cells or the endosperm in developing seeds to facilitate the absorption of nutrients from surrounding tissues.

hermaphrodite: an individual or structure bearing both male and female reproductive organs, e.g. a bisexual flower has both fertile stamens and carpels. In Greek mythology *Hermaphroditos* was the name of the handsome son of Hermes and Aphrodite, who became united with the nymph Salmacis in one body - half man, half woman.

heterospory (Greek: *heteros* = different, of another kind + *sporos* = germ, spore): production of two types of spores that differ with respect to their size and sexual differentiation. The larger, female spores are called *megaspores*, the smaller, male are called *microspores*.

hilum (Latin: *hilum* = little thing, trifle): in a seed the point of attachment of the funiculus or (if a funiculus is absent) of the placenta. The present meaning originates from the belief that the Latin word originally meant "something that adheres to a bean".

horsetail: common name for a member of the Sphenopyhta, a group of seedless, spore-producing vascular plants. Three hundred million years ago in the Carboniferous time period, lowland forest and swamps consisted of a great variety of spore-producing trees, the most prominent among them being relatives of clubmosses and horsetails. The members of the Sphenophyta are distinguished by their straight stems with branches or leaves arranged in regular whorls. Today, the sphenophytes consist of only one suriving genus, *Equisetum*, with about fifteen species worldwide.

hygrochastic: see hygrochasy.

hygrochasy (Greek: *hygros* = wet, moist + *chasis* = crack, gullet): production of hygroscopically dehiscing capsules that open only when wet (and often close again upon desiccation), e.g. the capsules of many succulent Aizoaceae.

hypocotyl (Greek: *hypo* = under, beneath + *cotyledon*): the axis of the embryo delimited by the radicle at one end and by the point of insertion of the cotyledons at the other. The name alludes to the position of the short stem below the cotyledons.

indusium (Greek: *endusis* = dress): in ferns a membranous outgrowth of the leaf epidermis covering a sorus (a cluster of sporangia).

inflorescence: part of a plant which bears a group of flowers; inflorescences can be a loose group of flowers (as in lilies) or highly condensed and differentiated structures resembling an individual flower as in the sunflower family (Asteraceae).

integument (Latin: *integumentum* = cover, blanket): layer enveloping the megasporangium (nucellus) of seed plants. In gymnosperms the nucellus is covered by one (cycads, conifers), two (*Ephedra, Welwitschia*) or three (*Gnetum*) integuments; angiosperms have either one or two integuments.

Jurassic: geological time period 142-206 million years ago.

lagenostome (Greek: *lagenos* = flask, bottle + *stoma* = mouth): the pollen-collection funnel-shaped structure at the apex of the ovules of the earliest seed plants (pteridosperms).

Lepidodendron (Greek: *lepis* = scale + *dendron* = tree): giant tree-like clubmoss relative from the Carboniferous time period. See also clubmoss.

loculus/locule (Latin: *loculus* = little place, diminutive of *locus* = place): the seed-bearing cavity of a carpel.

megagametophyte (Greek: *megas* = big + *gametes* = spouse + *phyton* = plant): female prothallus developed from the megaspore, eventually producing the female gametes (egg cells). The megagametophyte of the gymnosperms gives rise to the archegonia and provides the nutritious tissue of the seed, in angiosperms the homologue of the megagametophyte is the embryo sac.

megasporangium (Greek: *megas* = big + *sporos* = germ, spore + *angeion* = vessel,

small container) the organ of a sporophyte that produces the female megaspores. The term is usually applied to cryptogams, whereas the homologous structure in seed plants is called nucellus.
megaspore mother cell: a cell which gives rise to four haploid female spores during meiosis.
megaspore (Greek: *megas* = big + *sporos* = germ, spore): the larger spore formed by heterosporous plants which gives rise to a female gametophyte.
megasporocyte (Greek: *megas* = big + *sporos* = germ, spore + *kytos* = vessel): = megaspore mother cell.
megasporophyll (Greek: *megas* = big + *sporos* = germ, spore + *phyllon* = leaf): specialised fertile leaf producing megasporangia with female spores, e.g. the ovule-bearing carpel of the angiosperms.
meiosis (Greek: *meiosis* = diminution, reduction, from *meioun* = to diminish, to reduce): a type of nuclear division that results in daughter nuclei each containing half the number of chromosomes of the parent, i.e. chromosome number is reduced from diploid to haploid. Meiosis is completed after two subsequent cell divisions so that one diploid cell gives rise to four haploid ones.
mesocarp (Greek: *mesos* = middle + *karpos* = fruit): the fleshy middle layer of the fruit wall (pericarp).
microgametophyte (Greek: *megas* = big + *gametes* = spouse + *phyton* = plant): male prothallus developed from the microspore, eventually producing the male gametes (sperm). The microgametophyte of the cryptogams gives rise to special organs (antheridia) producing male gametes. In seed plants the microgametophyte is strongly reduced (represented by the germinated pollen grain) and no longer produces antheridia, only sperm nuclei.
micrometer (Greek: *mikros*): one millionth of a metre (a thousandth of a millimetre), abbreviated Ìm.
micropyle (Greek: *mikros* = small + *pyle* = gate): the opening of the integument(s) at the apex of the ovule or seed, usually acting as a passage for the pollen tube on its way to the embryo sac. The micropyle is formed by one or both integuments, the outer one producing the exostome, the inner one the endostome.
microsporangium (Greek: *mikros* = small + *sporos* = germ, spore + *angeion* = vessel, small container): the organ of a sporophyte that produces the male microspores. The term is usually applied to cryptogams, whereas the homologous structure in seed plants is called pollen sac.
microspore mother cell: a cell which gives rise to four haploid male spores (called pollen in seed plants) during meiosis.
microspore (Greek: *mikros* = small + *sporos* = germ, spore): the smaller spore formed by heterosporous plants which gives rise to a male gametophyte.
microsporocyte = microspore mother cell.
microsporophyll (Greek: *mikros* = small + *sporos* = germ, spore + *phyllon* = leaf): specialised fertile leaf-producing microsporangia with male spores, e.g. the pollen-bearing stamen of the angiosperms.
mitosis (Greek *mitos*, warp thread + Latin suffix: *-osis*): the typical process of cell division during which one cell divides into two genetically identical daughter cells (as opposed to meiosis during which the number of chromosomes in each daughter cell is reduced to half).
Monocots = Monocotyledons.
Monocotyledons (Greek: *monos* = one + cotyledon): one of the two major groups of the angiosperms distinguished by the presence of only one leaf (cotyledon) in the embryo. Other typical characters of the Monocotyledons are parallel leaf venation, floral organs usually in threes, scattered vascular bundles, a rudimentary primary root which is soon replaced by lateral adventitious roots (i.e. roots formed by the stem), and the lack of secondary thickening which is the reason why most Monocotyledons are herbaceous plants.
monoecious: see monoecy.
monoecy (Greek: *monos* = one + *oikos* = house): (1) the formation of male and female sexual organs on the same gametophyte (e.g. in many mosses and ferns), (2) in seed plants the formation of male and female flowers on the same individual.
morphology (Greek: *morphe* = shape, *logos* = word, speech): the study of form in the widest sense but mostly restricted to the external structure of an organism as opposed to anatomy which refers to the internal structure of an organism.
multiple fruit: a fruit that develops from an apocarpous gynoecium. Most modern authors apply this term to a fruit that is derived from more than one flower (compound fruit) and call fruits developing from apocarpous gynoecia "aggregate fruits". However, "aggregate fruit" is historically synonymous with "compound fruit", both defined as being composed of more than one flower (Spjut and Thieret 1989). Spjut and Thieret (1989) traced the confusion to Lindley (1832) who had reversed meanings for aggregate and multiple as defined by de Candolle (1813). English text books have generally adopted Lindley's errors, while non-English text books have followed de Candolle's (1813) definitions, or have employed other related terms. To avoid further confusion between aggregate and multiple, Spjut and Thieret (1989) recommended that the term compound fruit be adopted instead of aggregate fruit for fruits composed of more than one flower, and that the original and correct meaning for multiple fruit be maintained. The distinction between multiple and compound fruits was first made by Gärtner (1788), but made more clearly by Link (1798).
mycotrophic (Greek: *mycos* = fungus + *trophos* = feeder): referring to plants living in symbiosis with fungi.
myrmecochory (Greek: *myrmex* = ant + *chorein* = to wander, to disperse): dispersal of fruits and seeds by ants.
nucellus (Modern Latin: *nucellus* = little nut): the megasporangium of the seed plants.
nucleus (plural nuclei; Latin: *nuculeus* or *nucleus* = kernel, core, from *nucula* = little nut): compartment of a cell in which its genetic information is located.
nut: a dry, indehiscent and usually single-seeded fruit.
ornithochory (Greek: *ornis* = bird + *chorein* = to disperse, to wander): dispersal of fruits and seeds by birds.
ovary (Modern Latin: *ovarium* = a place or device containing eggs, from Latin: *ovum* = egg): the enlarged, usually lower portion of a pistil containing the ovules.
ovule (Modern Latin: *ovulum* = small egg): the integumented megasporangium of the seed plants.
Pangaea: the ancient supercontinent in which all continents of the earth were once united before they were separated by continental drift.
pappus (Latin: *pappus* = old man, from Greek: *pappos* = grandfather, old man, old man's beard): bristles, awns, hairs or scales that develop at the upper margin of the fruit of the Asteraceae, possibly representing a reduced calyx. A pappus is often an adaptation for wind dispersal of the fruit (cypsela), e.g. in dandelion (*Taraxacum officinale*), meadow salsify (*Tragopogon pratensis*).
parthenogenesis (Greek: parthenos = virgin + Greek: *genesis* = birth, beginning) form of asexual reproduction whereby an egg cell develops into an embryo without prior fertilization by a male gamete. Parthenogenesis is usually the result of an abnormal meiosis resulting in an egg nucleus with an unreduced number of chromosomes (e.g. in dandelion, *Taraxacum officinale*, Asteraceae).
perianth (Greek: *peri* = around; *anthos* = flower): the floral envelope that is clearly differentiated into calyx (outer perianth whorl) and corolla (inner perianth whorl).
pericarp (Modern Latin: *pericarpum*, from Greek: *peri* = around + *karpos* = fruit): the wall of the ovary at the fruiting stage. The pericarp can be homogeneous (as in berries) or differentiated into three layers (as in drupes) called epicarp, mesocarp, and endocarp.
perisperm (Greek: *peri* = around + *spermatos* = seed): a diploid storage tissue in angiosperm seeds, derived from the nucellus. The Greek name alludes to the embryo bending around the central storage nucellus as in the seeds of the Amaranth family (Amaranthaceae) and pink family (Caryophyllaceae).
Permian: geological time period 248-290 million years ago.
petal (Modern Latin: *petalum*, from Greek *petalon* = leaf): in flowers where the outer whorl of the perianth is different from the inner whorl the elements of the inner whorl of the floral envelope are addressed as petals. All the petals together form the often brightly coloured, showy corolla of a flower.
pistil (Latin: *pistillum* = pestle; alluding to the shape): an individual ovary with one or more styles and stigmas, composed of one or more carpels; introduced in 1700 by Tournefort, now mostly replaced by the gynoecium.
placenta (Modern Latin: *placenta* = flat cake, originally from Greek *plakoenta*, accusative of *plakoeis* = flat, related to *plax* = anything flat): a region within the ovary where the ovules are formed and remain attached (usually via a funiculus) to the parent plant until the seeds are mature. In botany the term was adopted from the similar structure to which the embryo is attached in animals and humans.
polar nuclei: the two nuclei in the central cell of the embryo sac of the angiosperms.
pollen chamber: a chamber at the apex of the nucellus of the ovules of many gymnosperms where the pollen grains end up and germinate.
pollen mother cell: the microspore mother cell of the seed plants, giving rise to four pollen grains during meiosis.
pollen sac: the microsporangium of the angiosperms. One anther typically bears four pollen sacs.
pollen tube: tube-like structure formed by the germinating pollen grain. In cycads and *Ginkgo* the pollen tube releases the motile sperm directly into the pollen chamber from where they swim to the archegonia. In conifers and angiosperms the pollen tube delivers the sperm nuclei straight to the egg cells.
pollen (Latin for *fine flour*): the microspores of the seed plants, able to germinate on or near the megasporangium to produce a very small and strongly simplified microgametophyte.
pollination drop: drop of liquid secreted by the micropyle of many gymnosperms as a means to collect pollen. The pollination drop is finally reabsorbed and the captured pollen sucked into the pollen chamber.
polyembryony (Greek: *poly-*, from *polys* = many + Latin: *embryo*): the formation of multiple embryos within one ovule. Polyembryony can be achieved either through formation of multiple zygotes (in gymnosperms through the fertilization of several archegonia in one ovule, e.g. Pinaceae), through one zygote giving rise to more than one embryo (e.g. through longitudinal division of the proembryo into four equal parts in *Pinus* spp., Pinaceae), or through the formation of adventive embryos from cells of the nucellus (e.g. in *Citrus* spp., Rutaceae).
pre-ovule: the ovule of the first seed plants (pteridosperms). In pre-ovules the integument did not yet form a complete envelope around the nucellus. Instead it consisted of several separate spreading lobes which surrounded the nucellus like a calyx, leaving its top visible. At its exposed apex the nucellus of pre-ovules possessed a funnel-shaped opening at the top, the lagenostome. The function of the lagenostome was to collect pollen from the air and pass them on into a pollen chamber underneath. This pollen chamber was situated right above the area where the megagametophyte produced its archegonia. In its centre at the bottom was the central column, a protrusion formed of nucellar tissue.
pre-pollen: fossil gymnospermous pollen believed to produce motile sperm, rather than effecting fertilization via a pollen tube.
progenesis (Latin: *pro-*: for [in the sense of supporting] + Greek: *genesis* = birth, beginning): condition in of an organism reaching sexual maturity while still in its juvenile stage (e.g. certain amphibians and insects).
prothallus (Greek: *pro* = before, in front + *thallos* = shoot): a small haploid (male, female or hermaphrodite) gametophyte. Prothalli are well developed in algae, mosses, ferns and fern allies and some gymnosperms. A prothallus develops from a haploid spore and produces either antheridia or archegonia or both. In angiosperms both male and female gametophytes are highly reduced (no more formation of antheridia and archegonia) with the pollen tube and embryo sac as the homologues of the microgametophyte and megagametophyte.
pseudocarp (Greek: *pseudos* = false, the lie, from *pseudein* = to lie + *karpos* = fruit): meaning "false fruit"; a term used in modern textbooks to denote a fruit in which not only the gynoecium but also other floral parts participate. The correct term for such a fruit is *anthocarp*.
pteridosperms (Greek: *pteris* = fern + *spermatos* = seed): fossil group of gymnosperms superficially resembling ferns; therefore also called "seed ferns".
radicle (Latin: *radicula* = small root): in an embryo the basal part of the hypocotyl-root axis producing the young seedling's first root system, often also called "embryonic root".
radiocarbon-dating: a method to establish the age of organic materials using their concentration of the radioactive carbon isotope C14. Radiocarbon dating, also called C14 dating, is believed to deliver reliable results up to an age of 50-60,000 years.
raphe (Greek: *raphe* = seam, suture, from *rhaptein* = to sew): area of the seed coat in which the continuation of the funicular vascular bundle runs from the hilum to the chalaza. The raphe is longest in anatropous seeds, shorter in moderately campylotropous seeds and entirely absent in strongly campylotropous and atropous seeds.
sarcotesta (Greek: *sarko* = flesh + Latin: *testa* = shell): a fleshy seed coat.
samara (Latin name for the fruit of the elm): a winged nut.
scanning electron microscope: a scientific instrument that produces images of very small objects with extremely high resolution by using electrons reflected from the surface of a specimen.
schizocarpic fruit (Modern Latin, from Greek: *skhizo-*, from *skhizein* = to split + *karpos* = fruit): one in which the carpels are partially to completely united at the time of pollination but separate at maturity into their carpellary constituents, sometimes further dividing into mericarps, each part functioning as a seed dispersal unit.
sclerotesta (Greek: *scleros* = hard + Latin: *testa* = shell): the inner hard layer of a sarcotestal seed. The term sclerotesta is more often applied to the stony layer of the seed coat of the gymnosperms than to comparable structures in the angiosperms.
seed: the organ of the seed plants (spermatophyta) that encloses the embryo together with a nutritious tissue within a protective seed coat. Seeds develop from integumented megasporangia (ovules), the defining organ of the seed plants.
seed plants: plants that produce seeds, see spermatophyta.

sepal (Modern Latin: *sepalum*, an invented word, perhaps a blend of Latin *petalum* and Greek: *skepe* = cover, blanket): in flowers where the outer whorl of the perianth is different from the inner whorl, the elements of the outer whorl are addressed as sepals. All the sepals together form the generally inconspicuous green calyx of a flower.

septum (plural septae; Latin: dissepimentum = wall, division): partition or diaphragm within an ovary.

serotiny (Latin: *serotinus* = coming late, from *sero* = at a late hour, from *serus* = late): late in developing or blooming; in the context of seeds serotiny refers to the condition in plants that maintain an "aerial seed bank" by holding on to their fruits and seeds long after they have matured. It is an adaptation to fire-prone habitats: the fruits of serotinous plants release their seeds after exposure to heat.

Sigillaria (Latin: *sigillum* = seal): giant tree-like clubmoss relative from the Carboniferous time period. See also clubmoss.

simple fruit: a fruit that develops from *one flower* with only *one pistil*; a pistil can be either a single carpel or several united carpels.

soil seed bank: the summary of all viable seeds present on and in the ground.

sorus (plural sori; Modern Latin *sorus*, from Greek *soros* = heap): a cluster or group of sporangia on the underside of the fronds of ferns.

sperm nucleus: the extremely reduced, non-motile male gamete of conifers and angiosperms.

spermatophyta/spermatophytes (Greek: *spermatos* = seed + *phyton* = plant): seed-producing plants. Group of plants characterised by the female gametophyte being developed and retained within an integumented megasporangium (ovule) which after fertilization of the egg cell develops into a seed. The spermatophytes comprise two major groups, the gymnosperms and the angiosperms.

spermatozoid (Greek: *sperma* = seed + *zoon* = animal): a motile male gamete.

sporangium (Greek: *sporos* = germ, spore + *angeion* = vessel, container): container with an outer cellular wall and a core of cells which give rise to spores.

spore: a cell serving asexual reproduction.

sporophyll (Greek: *sporos* = germ, spore + *phyllon* = leaf): fertile leaf carrying one or more sporangia. Heterosporous plants usually have specialised microsporophylls producing male (micro-)spores and megasporophylls producing female (mega-)spores.

sporophyte (*sporos* = germ, spore + *phyton* = plant): literally "the plant which produces the spores"; the diploid generation in the life cycle of plants which produces asexual, haploid spores that give rise to haploid gametophytes.

stamen (Latin: *stamen* = thread): the microsporophyll of the angiosperms, consisting of the sterile filament that carries the fertile anther at the apex; each anther bears four pollen sacs (microsporangia) containing the pollen grains (microspores).

stigma (Greek = spot, scar): the upper end of a carpel able to receive pollen grains; the stigma is usually elevated above the ovary by a style.

style (Greek: *stylos* = column, pillar): in angiosperms the narrow, elongated part of a carpel or pistil connecting stigma and ovary through which the pollen tubes grow down into the ovary.

suspensor (Latin: *suspendere* = to suspend; from *sub-* = under, from below + *pendere* = to hang): in seed plants an organ which develops from the zygote mainly by transverse cell divisions into a stalk-like organ, which carries the embryo at its tip.

suture (Latin: *sutura* = seam): a line marking a junction or fusion of organs, sometimes representing preformed lines of dehiscence along which, for example, a carpel of a dehiscent fruit opens; the dorsal suture of a carpel usually coincides with the central vascular bundle ("midrib") of the carpel, the ventral suture usually coincides with the line of fusion of the carpellary margins.

syncarpous gynoecium (Greek: *syn* = together + *karpos* = fruit and *gyne* = woman + *oikos* = house); literally "a joint women's house"; a gynoecium consisting of two or more joint carpels.

synergids (Greek: *synergos* = working together): the two cells flanking the egg cell in the embryo sac of the angiosperms; the synergids were obviously conceived as two cells working together to support the egg cell.

Triassic: geological time period 206-248 million years ago.

turgor pressure (Latin: *turgere* = to swell): the hydrostatic pressure inside a plant cell. The turgor pressure provides mechanical support and keeps plant cells turgid. When the turgor pressure drops, a plant starts wilting.

unitegmic ovules: ovules in which the nucellus is covered by only one integument.

zoochory (Greek: *zoon* = animal + *chorein* = to disperse, to wander): dispersal of fruits and seeds by animals.

zygote (Greek: *zygotos* = joined together): fertilised (diploid) egg cell.

BIBLIOGRAPHY

SEEDS & BOTANY

Arditti, J. & Abdul Karim Abdul Ghani (2000) Numerical and physical properties of orchid seeds and their biological implications (Tansley Review No. 110). *New Phytologist* 145: 367-421.

Armstrong, W.P. A nonprofit natural history textbook dedicated to little-known facts and trivia about natural history subjects. www.waynes-word.com

Baskin, C.C. & Baskin, J.M. (1998) *Seeds: ecology, biogeography, and evolution of dormancy and germination.* San Diego, CA: Academic Press.

Bailey, J. (editor) (1999) *The Penguin Dictionary of Plant Sciences.* Second edition (completely revised). Penguin Books, London, England.

Barthlott, W. & Ziegler, B. (1981) Mikromorphologie der Samenschalen als systematisches Merkmal bei Orchideen. *Berichte der Deutschen Botanischen Gesellschaft* 94: 267-27.

Bateman, R.M. & DiMichele, W.A. (1994) Heterospory: the most iterative key innovation in the evolutionary history of the plant kingdom. *Botanical Reviews* 69: 345-417.

Beattie, A.J. (1985) *The evolutionary ecology of ant-plant mutualism.* Cambridge University Press, Cambridge, UK.

Beck, C.B. (editor) (1988) *Origin and early evolution of gymnosperms.* Columbia University Press, New York, USA.

Beer, J.G. (1863) *Beiträge zur Morphologie und Biologie der Familie der Orchideen.* Wien.

Boesewinkel, F.D. and Bouman, F. (1984) The seed: structure. In: Johri, B.M (ed.): *Embryology of angiosperms*, pp. 567-610. Springer Verlag, Heidelberg.

Bond, W. & Sillingsby, P. (1984) Collapse of an ant-plant mutualism: the argentine ant (*Iridomyrmex humilis*) and myrmecochorous Proteaceae. *Ecology* 65: 1031-1037

Bouman, F. (1984) The ovule. In: Johri, J.B. (ed.): *Embryology of the angiosperms*, pp. 123-157. Springer Verlag, Berlin, Heidelberg, New York, Tokyo.

Bouman, F., Boesewinkel, D., Bregman, R., Devente, N. & Oostermeijer, G. (2000) *Verspreiding van zaden.* KNNV Uitgeverij, Utrecht.

Bresinsky, A. (1963) Bau, Entwicklungsgeschichte und Inhaltsstoffe der Elaiosomen. Studien zur myrmekochoren Verbreitung von Samen und Früchten. *Bibliotheca Botanica* 126: 1-54.

Brouwer, W. & Stählin, S.A. (1975) *Handbuch der Samenkunde, 2. Auflage.* DLG Verlag, Frankfurt.

Burrows, F.M. (1975) Wind-borne seed and fruit movement. *New Phytologist* 75: 405-418.

Camus, J.M., Jermy, A.C. & Thomas, B.A. (1991) *A World of Ferns.* Natural History Museum Publications, London.

Candolle, A.P. de (1813) *Théorie élémentaire de la botanique.* Déterville, Paris

Cook, C.D.K. (1990) Seed-dispersal of *Nymphoides peltata* (S.G. Gmelin) O. Kuntze (Menyanthaceae). *Aquatic Botany* 37: 325-340.

Corner, E.J.H. (1976) *The seeds of Dicotyledons.* Cambridge University Press, Cambridge, UK.

Fenner, M. & Thompson, K. (2005) *The ecology of seeds.* Cambridge University Press, Cambridge, UK.

Gaertner, J. (1788, 1790, 1791, 1792) *De fructibus et seminibus plantarum.* 4 Vols. Academiae Carolinae, Stuttgart.

Govaerts, R. (2001) How many species of seed plants are there?. *Taxon* 50, 1085-1090.

Gunn, C.R. & Dennis, J.V. (1999) *World guide to tropical drift seeds and fruits* (reprint of the 1976 edition). Krieger Publishing Company, Malabar, Florida, USA.

Harper, J.L., Lovell, P.H. & Moore, K.G. (1970) The shapes and sizes of seeds. *Annual Review of Ecology and Systematics* 1: 237-356.

Heywood, V.H. (editor) (1978) *Flowering Plants of the World.* Oxford University Press, Oxford, London, Melbourne.

Howe, H.F. & Smallwood, J. (1982) Ecology of seed dispersal. *Annual Review of Ecology and Systematics* 13: 201-228.

Jeffrey, C. (1989) *Biological Nomenclature.* 3rd Edition. Systematics Association, Edward Arnold, London

Jones, D.L. (2002) *Cycads of the world*, 2nd edition. Reed New Holland Publishers, Sydney, Auckland, London, Cape Town.

Kenrick, P. & Davis, P. (2004) *Fossil plants.* Natural History Museum, London, UK.

Kerner von Marilaun, A. & Oliver, F.W. (1903) *The Natural History of Plants.* Vol. II. The Gresham Publishing Company, London.

Kuijt, J. (1969) *The biology of parasitic flowering plants.* University of California Press, Berkeley, USA.

Leins, P. (2000) *Blüte und Frucht.* Schweizerbart'sche Verlagsbuchhandlung. Stuttgart, Berlin, 390 pp.

Lindley, J. (1832) *An introduction to botany.* Longman, Rees, Orme, Brown, Green & Longmans, London.

Link, H.F. (1798) *Philosophiae botanicae novae.* C. Dieterich, Göttingen.

Loewer, P. (2005) *Seeds - the definitive guide to growing, history and lore.* Timber Press, Portland, Cambridge, USA.

Mabberley, D.J. (1987) *The plant-book.* Cambridge University Press.

Mabberley, D.J. (1997) *The plant-book*, 2nd edition. Cambridge University Press.

Martin, A.C. (1946) The comparative internal morphology of seeds. *American Midland Naturalist* 36: 513-660.

Matlack, G.R. (1987) Diaspore size, shape, and fall behaviour in wind-dispersed plant species. *American Journal of Botany* 74: 1150-1160.

Netolitzky, F. (1926) *Anatomie der Angiospermen-Samen.* Berlin, Gebrüder Borntraeger (in: Linsbauer, K. (ed.): *Handbuch der Pflanzenanatomie* II. Abt., 2. Teil, Band 10).

Perry, E.L. IV & Dennis, J.V. (2003) *Sea-beans from the tropics – a collector's guide to sea-beans and other tropical drift on Atlantic shores.* Krieger Publishing Company, Malabar, Florida, USA.

Pijl, L. van der (1982) *Principles of dispersal in higher plants*, 3rd ed. Springer, Berlin, Heidelberg, New York.

Priestley, D.A. (1986) *Seed aging.* Cornell University, Ithaca, USA.

Rauh, W., Barthlott, W. & Ehler, N. (1975) Morphologie und Funktion der Testa staubförmiger Flugsamen. *Botanische Jahrbücher für Systematik, Pflanzengeschichte und Pflanzengeographie* 96: 353-374.

Raven, P.H., Evert, R.F. & Eichhorn, S.E. (1999) *Biology of plants.* W.H. Freeman, New York.

Rheede van Oudtshoorn, K. van & Rooyen, M.W. van (1999) *Dispersal biology of desert plants. Adaptations of Desert Organisms.* Springer-Verlag, Berlin.

Roth, I. (1977) *Fruits of angiosperms.* Gebrüder Borntraeger, Berlin & Stuttgart. 675 pp.

Sernander, R. (1906) *Entwurf einer Monographie der europäischen Myrmekochoren.* Kungliga Svenska Vetenskapsakademiens Handlingar 41: 1-410.

Shen-Miller, J. Mudgett, M.B., Schopf, J.W., Clarke, S. & Berger R. (1995) Exceptional seed longevity and robust growth: ancient Sacred Lotus from China. *American Journal of Botany* 82: 1367-1380.

Sorensen, A.E. (1986) Seed dispersal by adhesion. *Annual Review of Ecology and Systematics* 17: 443-463.

Spjut, R.W. (1994) A systematic treatment of fruit types. *Memoirs of the New York Botanic Garden* 70:1-182.

Spjut, R.W. & Thieret, J.W. (1989) Confusion between multiple and aggregate fruits. *Botanical Reviews* 55: 53-72.

Stearn, W.T. (1992) *Botanical Latin.* David & Charles, Devon.

Stützel, T. & Röwekamp, I. (1999) Bestäubungsbiologie bei Nacktsamern (Gymnospermen). In: Zizka, G. & Schneckenburger, S. (eds.) *Blütenökologie – faszinierendes Miteinander von Pflanzen und Tieren.* Kleine Senckenberg-Reihe Nr. 33 – Palmengarten Sonderheft Nr. 31.

Takhtajan (Takhtadzhyan), A. (editor) (1985, 1988, 1991, 1992, 1996, 2000) *Anatomia Seminum Comparativa.* Vol. 1-6. Nauka, Leningrad. [in Russian]

Ulbrich, E. (1919) *Deutsche Myrmekochoren; Beobachtungen über die Verbreitung heimischer Pflanzen durch Ameisen.* Fischer, Leipzig & Berlin.

Ulbrich, E. (1928) *Biologie der Früchte und Samen (Karpobiologie).* Springer, Berlin, Heidelberg, New York.

Van der Burgt (1997) Explosive seed dispersal of the rainforest tree Tetraberlinia morelina (Leguminosae - Caesalpiniodeae) in Gabon. *Journal of Ecology* 13: 145-151.

Wasserthal, L.T. (1997) The pollinators of the Malagasy star orchids *Angraecum sesquipedale*, *A. sororium* and *A. compactum* and the evolution of the extremely long spurs by pollinator shift. *Botanica Acta* 110: 1-17.

Werker, E. (1997) *Seed anatomy.* (Encyclopedia of plant anatomy, Vol. 10, Part 3: Spezieller Teil). Gebrüder Borntraeger, Berlin, Stuttgart.

ARCHITECTURE

Adam, H.C. (1999) *Karl Blossfeldt.* Prestel, Munich

Aldersly-Williams, H. (2003) *Zoomorphic – New Animal Architecture,* Laurence King, London

Davies, P.H (2002) *Photographing Flowers and Plants.* Collins & Brown, London

Diffey, T.J. (1993) Natural Beauty without metaphysics. Published in, *Landscape, natural beauty and the arts,* Cambridge University Press

Ede, S (2000) *Strange and Charmed.* Calouste Gulbenkian Foundation, London

Fairnington, M (2002) *Dead or Alive, Natural History Painting,* Black Dog Publishing, London

Frankel, F. (2002) *Envisioning Science, The Design and Craft of the Science Image.* MIT Press, Cambridge MA.

Gamwell, L. (2002) *Exploring the Invisible, Art Science and the Spiritual.* Princeton University Press, Princeton NJ.

Haeckel, E. (1904) *Art Forms in Nature.* Reprinted, 1998. Prestel, Munich

Kemp, M. (2000) *Visualizations, the Nature Book of Art and Science.* Oxford University Press, Oxford

Kesseler, R. (2001) *Pollinate.* Grizedale Arts and The Wordsworth Trust, Cumbria.

Lethaby, W.R (1922) *Form in Civilization, Collected essays on Art & Labour,* Oxford University Press, London

Martin, G & Laoec, R. (2002) *Macrophotography, learning from a master,* Abrams, New York

Portoghesi, P. (2000) *Nature and Architecture,* Skira Editore, Milan

Stafford, B.M. (1994) *Artful Science, Enlightenment, Entertainment and the Eclipse of the Visual Image.* MIT Press, Cambridge MA.

Stafford, B.M. (1996) *Good Looking, Essays on the Virtue of Images.* MIT Press, Cambridge MA.

Thomas, A. (1997) *The Beauty of Another Order, Photography in Science.* Yale University Press, New Haven

Thompson, D'Arcy (1917) *On Growth and Form,* reprinted, Cambridge University Press

Walter Lack, H. (2001) *Garden Eden.* Taschen, Koln

INDEX OF PLANTS ILLUSTRATED

Plant family	Latin name, and authority for name	Common British name	Country
Fabaceae	*Abrus precatorius* L.	rosary pea, paternoster pea, 174	native to the tropics worldwide
Malvaceae	*Abutilon angulatum* (Guill. & Perr.) Mast.	unknown, 205	collected in Botswana
Malvaceae	*Abutilon pictum* cv.	abutilon, 204	horticultural cultivar
Fabaceae	*Acacia cyclops* A. Cunn. & G. Don	coastal wattle, 174	collected in south-western Australia
Sapindaceae	*Acer pseudoplatanus* L.	sycamore maple, 95	native to Europe & Western Asia
Fabaceae	*Afzelia africana* Sm.	African mahogany, 150, 151	collected in Burkina Faso; native to tropical Africa
Caryophyllaceae	*Agrostemma githago* L.	corncockle, 206, 207	cult. Wakehurst Place; native to the Mediterranean
Malvaceae	*Alcea pallida* (Waldst. & Kit. ex Willd.) Waldst. & Kit.	pale hollyhock, 208, 209	native to central and south-eastern Europe
Alliaceae	*Allium ampeloprasum* L.	wild leak, 252, 253	native to Eurasia and North Africa
Asphodelaceae	*Aloe trachyticola* (H. Perrier) Reynolds	unknown, 67	collected in Madagascar; native to Madagascar
Thelypteridaceae	*Ampelopteris prolifera* (Retz.) Copel.	unknown, 27	cult. Kew; native to the Old World Tropics
Primulaceae	*Anagallis arvensis* L.	scarlet pimpernel, 162	cult. Wakehurst Place; native to Europe but widely naturalized
Fabaceae	*Anagyris foetida* L.	stinking bean trefoil, purging trefoil, 68	collected in Saudi Arabia; native to the Mediterranean
Plantaginaceae	*Anarrhinum orientale* Benth.	unknown, 240, 241	cult. Kew; native to the Mediterranean
Apiaceae	*Anethum graveolens* L.	dill, 83	commercial source; native to Central Asia
Orchidaceae	*Angraecum sesquipedale* Thouars	Malagasy comet orchid, 52, 53	native to Madagascar
Caryophyllaceae	*Arenaria franklinii* Douglas ex Hook.	Franklin's sandwort, 112, 113	collected in Idaho, USA; native to North America
Papaveraceae	*Argemone platyceras* Link & Otto	prickly poppy, cowboy's fried egg, chicalote, 210, 211	cult. Kew; native to the USA & Mexico
Cactaceae	*Ariocarpus retusus* Scheidw.	living rock cactus, 2, 3	commercial source; native to Mexico
Azollaceae	*Azolla filiculoides* Lam.	duckweed fern, 28	cult. Kew; native to all tropical and warm regions, naturalized in Europe
Cactaceae	*Aztekium ritteri* Boed.	aztec cactus, 76	commercial source; native to Mexico
Lecythidaceae	*Bertholletia excelsa* Bonpl.	Brazil nut, 89	native to Brazil
Acanthaceae	*Blepharis mitrata* C.B. Clarke	klapperbossie (Afrikaans), 110, 111	collected in South Africa; native to southern Africa
Cactaceae	*Blossfeldia liliputana* Werderm.	unknown, 77	collected in Bolivia; native to Argentina & Bolivia
Fabaceae	*Brachystegia utilis* Hutch. & Burtt Davy	njenje, 68	collected in Malawi; native to tropical Africa
Fabaceae	*Caesalpinia major* (Medik.) Dandy & Exell	yellow nickernut, 136	collected in Jamaica; native to the tropics
Portulacaceae	*Calandrinia eremaea* Ewar	twining purslane, 200, 201	collected in South Australia; native to Australia
Asteraceae	*Calendula officinalis* L.	pot marigold, 212, 213	cult. Kew; origin unknown
Raunculaceae	*Caltha palustris* L.	marsh marigold, 132	native to northern temperate regions
Apiaceae	*Carum carvi* L.	caraway, 82	native to the Mediterranean
Orobanchaceae	*Castilleja exserta* (A. Heller) T.I. Chuang & Heckard ssp. *latifolia* (S. Watson) T.I. Chuang & Heckard	purple owl's clover, exserted Indian paintbrush, 1	collected in California, USA; native to North America
Orobanchaceae	*Castilleja flava* S. Watson	yellow paintbrush, 65	collected in Idaho, USA; native to North America
Pinaceae	*Cedrus atlantica* (Endl.) Manetti ex Carrière	Atlas cedar, 46	cult. Kew; native to North Africa
Asteraceae	*Centaurea cyanus* L.	cornflower, 145	collected in the UK; native to Eurasia
Asteraceae	*Centaurea montana* L.	mountain bluet, 47	native to Europe
Fabaceae	*Centrolobium microchaete* (Mart. ex Benth.) Lima ex G. P. Lewis	tarara amarilla, 94	native to South America
Aizoaceae	*Cephalophyllum loreum* (L.) Schwant.	unknown, 133	native to the Cape Province, South Africa
Apocynaceae	*Cerbera manghas* L.	pink-eyed cerbera, 134	native from the Seychelles to the Pacific
Asteraceae	*Cichorium intybus* L.	chicory, 214, 215	collected in the UK; native to Eurasia
Orobanchaceae	*Cistanche tubulosa* (Schenk) Hook. f.	desert hyacinth, 120	collected in its native Saudi Arabia
Ranunculaceae	*Clematis vitalba* L.	traveller's joy, 107	collected in the UK; native to Eurasia & North Africa
Capparaceae	*Cleome gynandra* L.	cat whiskers, African spiderflower, 20	collected in Burkina Faso; native to the Old World tropics
Capparaceae	*Cleome* L. sp.	spiderflower, 178	collected in Madagascar
Gyrostemonaceae	*Codonocarpus cotinifolius* (Desf.) F.Muell.	desert poplar, native poplar (Australia), 152	collected in Western Australia; native to Australia
Combretaceae	*Combretum zeyheri* Sond.	large-fruited Bushwillow, Raasblaar, 78	native to southern Africa
Ranunculaceae	*Consolida orientalis* (M. Gay ex Des Moul.) Schrödinger (syn. *Delphinium orientale* M. Gay ex Des Moul.)	larkspur, 99	cult. Kew; native to southern Europe
Apiaceae	*Cuminum cyminum* L.	cumin, 83	commercial source; native to Mediterranean
Cycadaceae	*Cycas megacarpa* K.D. Hill	unknown, 36	native to Queensland
Cycadaceae	*Cycas revoluta* Thunb.	sago palm, 37	native to Japan
Primulaceae	*Cyclamen graecum* Link	Greek snowbread, 85	cult. Kew, native to Greece, Cyprus and Turkey
Plantaginaceae	*Cymbalaria muralis* P. Gaertn., B. Mey. & Scherb.	ivy-leafed toadflax, 143	collected in the UK; native to Europe and North America
Orchidaceae	*Dactylorhiza fuchsii* (Druce) Soó	common spotted orchid, 128, 129	collected in the UK; native to Europe
Alismataceae	*Damasonium alisma* Mill.	starfruit, 162	collected in the UK; native to Europe
Sarraceniaceae	*Darlingtonia californica* Torr.	California pitcher plant, cobra lily, 102	native to the western USA
Ranunculaceae	*Delphinium peregrinum* L.	larkspur, 98, 182	cult. Kew; native to central and eastern Mediterranean
Ranunculaceae	*Delphinium requienii* DC.	larkspur, 99	native to southern France, Corsica and Sardinia
Plantaginaceae	*Digitalis ferruginea* L.	rusty foxglove, 125	collected in the UK; native to southern Europe to western Asia
Plantaginaceae	*Digitalis lutea* L.	yellow foxglove, 124	collected in Belgium; native to western and central Europe
Plantaginaceae	*Digitalis purpurea* L.	purple foxglove, 12	collected in the UK; native to south-west and western-central Europe
Dipterocarpaceae	*Dipterocarpus cornutus* Dyer	keruing gombang (Malay), 91	native to South-East Asia
Dipterocarpaceae	*Dipterocarpus costulatus* Slooten	keruing bajan (Malay), 90	native to South-East Asia
Dipterocarpaceae	*Dipterocarpus grandiflorus* (Blanco) Blanco	keruing belimbing (Malay), 91	native to South-East Asia
Winteraceae	*Drimys winteri* J.R. Forst. & G. Forst.	Winter's bark tree, 44, 45	cult. Wakehurst Place; native from Mexcio to Terra el Fuego
Droseraceae	*Drosera capillaris* Poir.	pink sundew, 234, 235	native to the eastern USA
Droseraceae	*Drosera cistiflora* L.	cistus-flowered sundew, 116	native to South Africa, Western Cape
Droseraceae	*Drosera natalensis* Diels	Natal sundew, 235	native to South Africa, Mozambique & Madagascar
Droseraceae	*Drosophyllum lusitanicum* (L.) Link	Portuguese sundew, dewy pine, 244, 245	cult. Kew; native to Spain & Portugal
Dipterocarpaceae	*Dryobalanops aromatica* Gaertn.	kapur (Malay), 90	native to South-East Asia
Dryopteridaceae	*Dryopteris filix-mas* (L.) Schott	male fern, 24, 25	collected in the UK; native to northern temperate regions
Arecaceae	*Dypsis scottiana* (Becc.) Beentje & J. Dransf.	Raosy palm, 68	collected in its native Madagascar
Cactaceae	*Echinocereus laui* G. Frank	Lau's hedgehog cactus, 196	native to Mexico
Boraginaceae	*Echium vulgare* L.	viper's bugloss, 224, 225	collected in the UK; native to Eurasia
Zamiaceae	*Encephalartos inopinus* R.A. Dyer	unknown, 180	native to South Africa
Zamiaceae	*Encephalartos laevifolius* Stapf & Burtt Davy	unknown, 35	native to South Africa & Swaziland
Zamiaceae	*Encephalartos longifolius* (Jacq.) Lehm.	unknown, 148, 149	native to South Africa
Fabaceae	*Entada gigas* (L.) Fawc. & Rendle	sea-heart, 136	native to tropical America & Africa
Ephedraceae	*Ephedra equisetina* Bunge	bluestem joint fir, Ma huang, 38	cult. Kew; native to China & Japan
Onagraceae	*Epilobium angustifolium* L.	rosebay willowherb, fireweed, 108, 109	collected in the UK; native to northern temperate regions
Ericaceae	*Erica cinerea* L.	bell heather, 4	collected in the UK; native to Europe
Ericaceae	*Erica* x *darleyensis* 'Lena'	Darley heath, 5	horticultural cultivar
Euphorbiaceae	*Euphorbia damarana* L.C. Leach	milkbush, 175	collected in Namibia; native to the Namib Desert (Damaraland)
Euphorbiaceae	*Euphorbia epithymoides* L.	cushion spurge, 139	popular garden plant; native to central and eastern Europe
Euphorbiaceae	*Euphorbia helioscopia* L.	sun spurge, madwoman's milk, 138	collected in the UK; native to Eurasia & northern Africa
Euphorbiaceae	*Euphorbia* L. sp. LEB 390	spurge, 202	collected in Lebanon
Euphorbiaceae	*Euphorbia peplus* L.	petty spurge, 203	collected in the UK; native to Europe
Cupressaceae	*Fitzroya cupressoides* (Molina) I.M. Johnst.	Patagonian cypress, 30	monotypic genus from southern Chile and southern Argentina
Liliaceae	*Fritillaria meleagris* L.	snake's head fritillary, 250, 251	collected in the UK; native to Europe & Western Asia; naturalized in Britain

Family	Scientific name	Common name, pages	Notes
Rubiaceae	*Gardenia thunbergia* Hiern	wild gardenia, 147	native to South Africa
Ginkgoaceae	*Ginkgo biloba* L.	ginkgo, maidenhair tree, 38	native to Eastern China
Molluginaceae	*Glinus lotoides* L.	lotus sweetjuice, glinus, 62	collected in Burkina Faso; widely distributed in warm and tropical regions
Aizoaceae	*Glottiphyllum oligocarpum* L. Bolus	unknown, 189	native to the Western Cape, South Africa
Boraginaceae	*Hackelia* Opiz sp.	stickseed, 157	collected in Alberta, Canada
Pedaliaceae	*Harpagophytum procumbens* DC. ex Meisn.	devil's claw, 150	native to southern Africa & Madagascar
Apiaceae	*Hasselquistia aphrodytium* L.	hogweed, ??	native to northern temperate regions
Malvaceae	*Hermannia* L. sp.	unknown, 176	collected in the Limpopo Province, South Africa
Malvaceae	*Hermannia muricata* Eckl. & Zeyh.	unknown, 163	collected in the Western Cape, South Africa
Saxifragaceae	*Heuchera rubescens* Torr.	pink alumroot, pink heuchera, 163	collected in Utah/Wyoming, USA; native to North America
Asteraceae	*Hieracium pilosella* L.	mouse-ear hawkweed, 72, 73	collected in the UK; native to Eurasia & North America
Euphorbiaceae	*Hura crepitans* L.	sandbox tree, 140	native to South America & the Caribbean
Schisandraceae	*Illicium verum* Hook. f.	star anise, 88	origin: Southern China & Vietnam (only known in cultivation)
Balsaminaceae	*Impatiens glandulifera* Royle	Himalayan balsam, 141	collected in the UK; native to the Himalaya
Convolvulaceae	*Ipomoea pauciflora* M. Martens & Galeotti	unknown, 69	collected in its native Mexico
Hyacinthaceae	*Lachenalia lutea* Sims	unknown, 248	collected in its native South Africa
Hyacinthaceae	*Lachenalia peersii* Marloth ex W.F. Barker	bekkies (Afrikaans), 249	native to the Cape Province, South Africa
Orobanchaceae	*Lamourouxia viscosa* Kunth	unknown, 18	collected in its native Mexico
Boraginaceae	*Lappula spinocarpos* (Forssk.) Asch. ex Kuntze	unknown, 256	collected in Arabia
Fabaceae	*Lathyrus clymenum* L. (syn. *L. articulatus* L.)	jointed pea, 140	native to the Mediterranean
Asteraceae	*Leucochrysum molle* (DC.) Paul G. Wilson	hoary sunray, golden paper daisy, 104, 105	collected in its native Australia
Magnoliaceae	*Liriodendron tulipifera* L.	tulip tree, 42	cult. Kew; native to Eastern North America
Arecaceae	*Lodoicea maldivica* Pers.	Seychelles nut, coco-de-mer, 23, 137	native to the Seychelles
Cactaceae	*Lophophora williamsii* (Lem. ex Salm-Dyck) J.M. Coult.	peyote, 216, 217	cult. Kew; native to the Chihuahuan Desert of Texas and northern Mexico
Caryophyllaceae	*Lychnis flos-cuculi* L.	ragged robin, 222, 223	collected in the UK; native to Eurasia
Proteaceae	*Macadamia integrifolia* Maiden & Betche	smooth-shelled Macadamia nut, 79	native to Queensland, Australia
Cactaceae	*Mammillaria theresae* Cutak	unknown, 163	native to Mexico
Marsileaceae	*Marsilea quadrifolia* L.	European waterclover, 28	cult. Kew; native to Europe
Cactaceae	*Melocactus neryi* K. Schum.	turk's-cap cactus, 218, 219	cult. Kew; native to South America
Cactaceae	*Melocactus zehntneri* (Britton & Rose) Luetzelb.	turk's-cap cactus, 58	cult. Kew; native to north-eastern Brazil
Convolvulaceae	*Merremia discoidesperma* (Donn. Sm.) O'Donell	Mary's bean, crucifixion bean, 136	collected in the New Hebrides, Scotland; native to Central America
Fabaceae	*Mora megistosperma* (Pittier) Britton & Rose (syn. *Mora oleifera* (Triana ex Hemsl.) Ducke)	nato mangrove, 70	native to tropical America
Fabaceae	*Mucuna urens* (L.) Medik.	true sea bean, hamburger bean, 136	collected in Panama; native to Central & South America, Caribbean and Hawai'i
Melanthiaceae	*Narthecium ossifragum* (L.) Huds.	bog asphodel, 118, 119	collected in the UK; native to Europe
Nelumbonaceae	*Nelumbo nucifera* Gaertn.	sacred lotus, Egyptian bean, 166, 167	native to Asia, New Guinea and Australia
Plantaginaceae	*Nemesia versicolor* E. Mey. ex Benth.	leeubekkie (Afrikaans), 168	collected in the Eastern Cape, South Africa; native to South Africa
Ranunculaceae	*Nigella damascena* L.	love-in-a-mist, 86, 87	cult. Kew, origin Mediterranean
Menyanthaceae	*Nymphoides peltata* (S.G. Gmel.) Kuntze	yellow floatingheart, 130, 131	collected in the UK; native to Eurasia
Ochnaceae	*Ochna natalitia* (Meisn.) Walp.	coast boxwood, Natal plane, 84	native to southern Africa
Orchidaceae	*Ophrys sphegodes* Mill.	early spider orchid, 23	collected in the UK; native to Europe
Apocynaceae	*Orbea semota* (N.E. Brown) L.C. Leach	unknown, 47	native to East Africa
Hyacinthaceae	*Ornithogalum dubium* Houtt.	yellow star-of-Bethlehem, 164, 165	collected in the Western Cape, South Africa; native to South Africa
Orobanchaceae	*Orobanche* L. sp.	broomrape, 146	collected in Greece
Orobanchaceae	*Orthocarpus luteus* Nutt.	yellow owl's clover, 122	collected in Alberta, Canada; native to North America
Paeoniaceae	*Paeonia cambessedesii* Willk.	peonia, 146	cult. Kew; native to the Balearic Islands (Cabrera, Mallorca, Minorca)
Parnassiaceae	*Parnassia fimbriata* K.D. Koenig	fringed grass-of-Parnassus, 171	collected in Oregon, USA; native to North America
Cactaceae	*Parodia magnifica* (F. Ritter) F.H. Brandt	unknown, 220, 221	commercial source; native to origin Brazil
Passifloraceae	*Passiflora caerulea* L.	blue passionflower, 29, 56	cult. Kew; native from Brazil to Argentina
Passifloraceae	*Passiflora* cv. 'Lady Margaret'	passionflower 'Lady Margaret', 29	cult. Kew
Paulowniaceae	*Paulownia tomentosa* (Thunb.) Steud	princess tree, 14	collected in the UK; native to China
Caryophyllaceae	*Petrorhagia nanteuilii* (Burnat) P.W. Ball & Heyw.	childing pink, 114, 115	cult. Wakehurst Place; native to Europe
Boraginaceae	*Pholistoma auritum* (Lindl.) Lilja ex Lindbl. var. *auritum*	fiesta flower, 186	collected in California, USA; native to western North America
Apiaceae	*Pimpinella anisum* L.	anise, 82	commercial source; native from Greece to Egypt
Pinaceae	*Pinus lambertiana* Douglas	sugar pine, 34, 40	collected in California, USA
Pinaceae	*Pinus tecunumanii* Eguiluz & J.P. Perry	piño rojo, 41	native to Central America
Bignoniaceae	*Pithecoctenium crucigerum* (L.) A.H. Gentry	monkey pod, 96	collected in Mexico; native to tropical America
Plantaginaceae	*Plantago media* L.	hoary plantain, 66	collected in the UK; native to Eurasia
Podocarpaceae	*Podocarpus lawrencei* Hook. f.	mountain plum pine, 147	cult. Kew; native to South Australia, & Tasmania
Polygalaceae	*Polygala arenaria* Oliv.	sand milkwort, 153	collected in Burkina Faso; native to tropical Africa
Zygophyllaceae	*Porlieria chilensis* I.M. Johnst.	guayacán, 69	collected in its native Chile
Aizoaceae	*Prenia tetragona* (Thunb.) Gerbaulet	unknown, 185	collected in the Cape Province, South Africa
Martyniaceae	*Proboscidea louisianica* (Mill.) Thell.	common unicorn plant, pale devil's claw, 158	native to the southern USA
Lythraceae	*Punica granatum* L.	pomegranate, 74, 75	native to Asia (Middle East to Himalaya)
Ericaceae	*Pyrola secunda* L.	one-sided pyrola, 127	commercial source; native to North America
Rhizophoraceae	*Rhizophora mangle* L.	red mangrove, 135	widespread in the south-west Pacific and the Americas
Salviniaceae	*Salvinia* Ség. sp.	floating fern, 28	common in tropical and warm-temperate regions
Saxifragaceae	*Saxifraga umbrosa* L.	London pride, 10	endemic to the Western and Central Pyrenees
Lamiaceae	*Scutellaria columnae* All.	skullcap, 50, 51	collected in Greece; native to the Mediterranean
Lamiaceae	*Scutellaria galericulata* L.	common skullcap, 198, 199	collected in the UK; native to northern temperate regions
Lamiaceae	*Scutellaria orientalis* L.	oriental skullcap, 80	cult. Kew; native to southern Europe
Crassulaceae	*Sedum album* L.	white stonecrop, 226, 227	collected in the UK; native to Europe
Poaceae	*Setaria viridis* (L.) P. Beauv.	green bristlegrass, 54, 55	collected in the UK; native to Eurasia
Dipterocarpaceae	*Shorea macrophylla* (De Vr.) Ashton	engkabang jantong (Malay), 91	native to Borneo
Caryophyllaceae	*Silene dioica* (L.) Clairv.	red campion, 254, 255	collected in the UK; native to Europe
Caryophyllaceae	*Silene gallica* L.	small-flowered catchfly, 228, 229, 264	collected in the UK; native to Europe
Caryophyllaceae	*Silene maritima* With.	sea campion, 230, 231	cult. Wakehurst Place; native to Europe
Caryophyllaceae	*Spergularia media* (L.) C. Presl.	greater sea-spurrey, greater sand-spurrey, 100	collected in Belgium; native to coastal Europe, the Mediterranean & Asia
Caryophyllaceae	*Spergularia rupicola* Lebel ex Le Jolis	rock sea-spurrey, 48, 49	cult. Wakehurst Place; native to the UK & France
Rubiaceae	*Spermacoce senensis* (Klotzsch) Hiern	unknown, 69	collected in Malawi; native to Africa
Orchidaceae	*Stanhopea* 'Assidensis' [*S. tigrina* x *S. wardii*]	stanhopea 'Assidensis', 233	artificial hybrid
Orchidaceae	*Stanhopea tigrina* Bateman ex Lindl.	tiger-like stanhopea, 232	cult. Kew; native from Eastern Mexico to Brazil
Caryophyllaceae	*Stellaria holostea* L.	greater stitchwort, 6, 7	collected in the UK; native to Europe & the Mediterranean
Caryophyllaceae	*Stellaria pungens* Brogn.	prickly starwort, 192	collected in New South Wales, Australia; native to Australia
Strelitziaceae	*Strelitzia reginae* Aiton	bird-of-paradise flower, 174	native to South Africa
Boraginaceae	*Symphytum officinale* L.	comfrey, 236, 237	native to Europe & the Mediterranean
Taxaceae	*Taxus baccata* L. 'Fructo luteo'	English yew, variety with yellow fruits, 147	cult. Kew
Gyrostemonaceae	*Tersonia cyathiflora* (Fenzl) A.S. George ex J.W. Green	button creeper, 152	collected in its native Western Australia.
Rutaceae	*Thamnosma africanum* Engl.	unknown, 8	collected in the Limpopo Province, South Africa; native to Africa
Brassicaceae	*Thlaspi arvense* L.	field penny-cress, 238, 239	collected in the UK; native to Europe
Malvaceae	*Tilia cordata* Mill.	small-leaved lime tree, 92, 93	collected in the UK; native to Europe
Saxifragaceae	*Tolmiea menziesii* (Hook.) Torr. & A. Gray	piggyback plant, mother-of-thousands, 190	collected in Oregon, USA; native to western North America
Boraginaceae	*Trichodesma africanum* (L.) Lehm.	unknown, 194, 195	collected in Saudi Arabia; native to North Africa, Arabian Peninsular to Iran
Arecaceae	*Trithrinax brasiliensis* Mart.	Brazilian needle-palm, saho palm, 71	native to Brazil
Apocynaceae	*Tweedia caerulea* D. Don	southern star, 106	native to temperate South America
Pedaliaceae	*Uncarina* (Baill.) Stapf sp.	unknown, 160, 161	native to Madagascar
Scrophulariaceae	*Verbascum sinuatum* L.	wavyleaf mullein, 246, 247	native to the Mediterranean
Scrophulariaceae	*Verbascum thapsus* L.	common mullein, 242, 243	native to Eurasia
Welwitschiaceae	*Welwitschia mirabilis* Hook. f.	tree tumbo, tumboa, 39	native to the desert of South-West Africa (Angola, Namibia)

FOOTNOTES

1 Hence their scientific name, derived from Greek: *kryptos* = hidden + *gamein* = to marry, to copulate; referring to "those who copulate in secrecy".

2 The expression *alternation of generations* refers to the regular alternation between a spore-producing generation, called sporophyte, and a generation that produces gametes (i.e. egg cells and sperm cells), called *gametophyte*.

3 In all fairness, it needs to be said that a moss sporophyte is just not able to grow any bigger. The reason for this lies in their primitive plumbing. Mosses do not yet possess proper roots and a network of vessels to transport water efficiently from the soil up into the plant body. If they grew taller, their water supply system simply would not be able to replace all of the water that evaporates from their leaves.

4 Why so many gymnosperms produce a lot more archegonia than they need to obtain an embryo is uncertain. Most likely, this apparent waste of material and energy is a primitive characteristic inherited from their non-seed bearing ancestors.

5 To be fair it has to be said that the anthophytes also include some gymnosperms, the extinct cycad-like *Bennettitales*, the closely related *Pentoxylon* and the present-day Gnetales order (comprising the three genera *Ephedra*, *Gnetum*, and *Welwitschia*).

6 Humphrey Bogart (Rick) to Claude Rains (Louis) at the close of the movie *Casablanca* (1942).

7 So called because of its slightly fuzzy leaves that feel like cardboard when rubbed.

8 In the Greek mythology *Hermaphroditos* was the name of the handsome son of Hermes and Aphrodite, who became united with the nymph Salmacis in one body – half man, half woman.

9 Greek: *syn* = together + *karpos* = fruit and *gyne* = woman + *oikos* = house; literally "a joint women's house" The counterpart to the syncarpous gynoecium with joined carpels is the the previously described apocarpous gynoecium with separate carpels [Greek: *apo* = being apart from + *karpos* = fruit]. The term *gynoecium* on its own refers to the summary of all carpels in a flower, irrespective of whether they are united or separate.

10 In the most primitive angiosperms such as the star anise family (Illiciaceae), each ovule can initiate more than one megaspore mother cell (e.g. 3-8 in *Illicium mexicanum*), although ultimately only one functional megaspore usually survives.

11 Recent research has shown that the typical seven-celled/eight-nucleate embryo sac, present in 99% of all angiosperms, could have originated from a duplication of an even simpler, four-nucleate/four-celled embryo sac as it is found in some of the most primitive members of the group, for example the water lilies (Nymphaeaceae) and the star anise family (Illiciaceae).

12 The few primitive extant angiosperms with a four-celled embryo sac have only one nucleus in the central cell and therefore produce a diploid, rather than a triploid, endosperm.

13 Sebastian Smee, exhibition review of the work photographer Giles Revell and painter Mark Fairnington at the Natural History Museum, London. *Daily Telegraph*, 1 May 2004

14 Dr. H. Walter Lack, writing in the introduction to *Garden Eden, Masterpieces of Botanical Illustration*. Cologne, Taschen, 2001

15 Gisborn, M. from "Excursions into (Post-)Humanist Painting," *Dead and Alive, Natural History Painting of Mark Fairnington*, London, Black Dog Publishing, 2002

16 Diffey, Dr. T., *Natural beauty without metaphysics*, published in *Landscape, natural beauty and the arts*, Cambridge University Press 1993

ACKNOWLEDGMENTS

We are especially grateful to the former and present Heads of the SCD, Roger Smith and Paul Smith, and line manager, John Dickie at the Millennium Seed Bank, for allowing WS to participate in this most exciting project. Their support and encouragement have been invaluable. At the Jodrell Laboratory we thank the Micromorphology Section, especially Paula Rudall and Chrissie Prychid, for helping us to use the Scanning Electron Microscope. At the SCD, the kind support of the members of the Curation Section was instrumental in sourcing most of the material we needed for our work, most notably Janet Terry and James Wood (now at the Royal Tasmanian Botanical Gardens); some of the most exciting specimens go back to recommendations made by James. We owe special gratitude to Elly Vaes, WS's talented assistant, for spotting many of the most interesting seed samples in the vast collection of the Millennium Seed Bank, for providing us with some of her digital images and for preparing the excellent diagrams. We also fondly remember Consuelo "Chelo" Belda Revert, a student from Spain (sponsored by the Leonardo da Vinci Programme) whom we thank for her help in the hunt for the exceptional. We are deeply indebted to Sir Peter Crane, Richard Bateman and Paula Rudall for their very thorough review of the manuscript and to Sheila de Vallée for editing the text. We are also grateful to all our friends and colleagues who kindly read some or all of the text and gave us their opinions: Steve Alton (who also provided us with seeds of *Drosera cistiflora*, *D. capillaris* and *Darlingtonia californica*), Erica Bower, Matthew Daws, John Dickie, Madeline Harley, Petra Hoffmann, Ilse Kranner, Hugh Pritchard, Tom Robinson, Richard Spjut, Elly Vaes and James Wood. Furthermore, we both wish to thank many colleagues at Kew who kindly offered their expertise and time to help us answer difficult questions and provide us with important material, in particular Aljos Farjon (Herbarium), Lucy Blythe (Foundation & Friends), Martin Cheek (Herbarium), David Cooke (HPE), Phil Cribb (Herbarium), Gina Fullerlove (Publications), Anne Griffin (Library), Phil Griffiths (Horticulture & Public Education – HPE), Tony Hall (retired, formerly HPE, now Kew Research Associate), Chris Haysom (HPE), Petra Hoffmann (Herbarium), Kathy King (HPE), Gwilym Lewis, (Herbarium), Riikka Lundahl (Directorate), Andy McRobb (Media Resources), Mark Nesbitt (Centre for Economic Botany), Grace Prendergast (Micropropagation, who kindly also provided us with orchid seeds), Chrissie Prychid (Jodrell Laboratory), Dave Roberts (Herbarium), Susan Runyard (Public Relations), Ruth Ryder (Jodrell Laboratory), Brian Schrire (Herbarium), Julia Steele (Centre for Economic Botany), Nigel Taylor (Curator, HPE) and Daniela Zappi (Herbarium). At Central Saint Martins College of Art & Design, we thank Kathryn Hearn (Course Director Ceramic Design,) Sylvia Backemeyer (Curator, Museum & Contemporary Collection). Paul Holt (Samphire Hoe), Alex Barclay (NESTA). We thank Simon Andrew Irvin (New Zealand) for lending us the fruit of Uncarina. Outside the United Kingdom we thank Wilhelm Barthlott (University of Bonn, Germany) for technical advice, Ernst van Jaarsveld and Anthony Hitchcock (Kirstenbosch Botanic Garden, Cape Town, South Africa), and Johan Hurter (Lowveld National Botanical Garden, Nelspruit, South Africa), for their time, hospitality and permission to photograph plants in their amazing collections, and, in Greece, Vassili and Jo Mouha. Finally, we are grateful to our close friends and families for their patient support, Emma Lochner-Stuppy for volunteering to be the first reader of the manuscript – her everlasting patience and love was most precious during the seven months of this project – and her mother Ronelle Lochner, who so patiently supported the hunt for photographs during a recent trip to South Africa. We thank John and Sandy Chubb, Agalis Manessi and Marco Kesseler for always being there.

Rob Kesseler & Wolfgang Stuppy

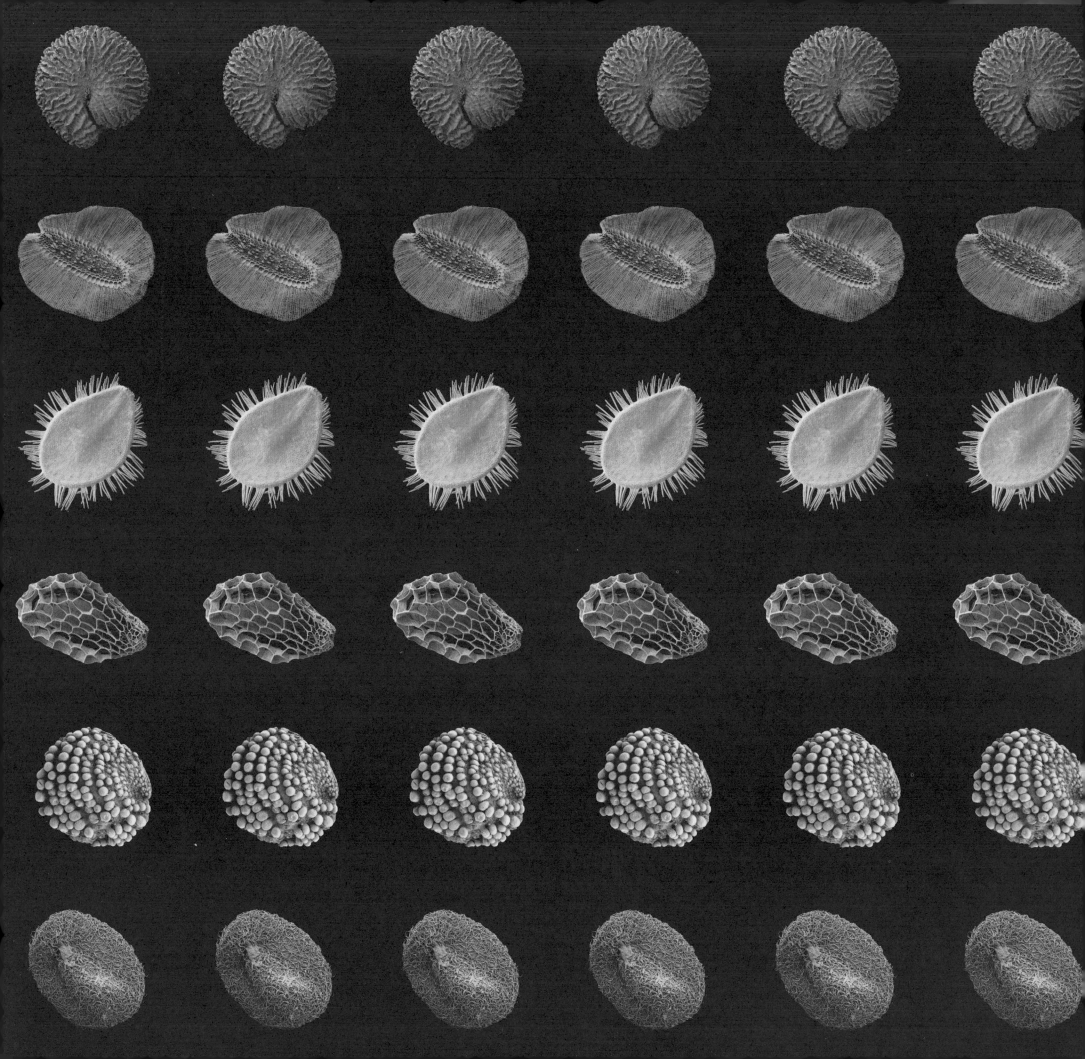